In the
SANCTUARY
of
OUTCASTS

In the
SANCTUARY
of
OUTCASTS

a memoir

Neil White

wm

WILLIAM MORROW
An Imprint of HarperCollins*Publishers*

Truth, Rumor, Leprosy, and Privacy

In the Sanctuary of Outcasts is a memoir. It is a true account of my experience as a federal inmate incarcerated at the Federal Medical Center in Carville, Louisiana. I witnessed an unprecedented convergence of cultures at Carville—federal inmates and prison guards were thrown together with leprosy patients, public health workers, and an ancient order of nuns. Early in my prison sentence, I fantasized that my incarceration served as a rare opportunity for participatory journalism. I interviewed hundreds of inmates, and all the Carville residents who would talk to me, carrying a notebook and pen wherever I went. I was discharged from Carville with a library of details recorded moments after the events and conversations occurred.

Part of what makes institutions—penal, mental, or medical—so intriguing is that all are plagued by lore, innuendo, and rumor. Whenever possible, facts have been checked to validate the stories told to me, but many details of the narratives, events, and personal histories recounted are not verifiable. These recollections are set forth as they were recounted, and I have included questions and conflicting narrative in the manuscript.

A diagnosis of leprosy can destroy lives. Until the 1970s, many newly diagnosed patients changed their names rather than risk exposing their families to the stigma and shame surrounding the disease. Some families, to this day, carry the secret that a relative had contracted leprosy. Out of respect for the Carville residents (and for their families) I have disguised their identities. Although several photographs of Carville residents have been included in the book—those who regularly sat for visiting photographers—I have disguised their names, family connections, and hometowns.

I have also changed the names of other individuals—and modified identifying features, including physical descriptions and occupations, of certain other individuals—in order to preserve their anonymity. The goal in all cases was to protect people's privacy without damaging the integrity of the story.

HarperCollins books may be purchased for educational, business, or sales promotional use. For information please write: Special Markets Department, HarperCollins Publishers, 10 East 53rd Street, New York, NY 10022.

FIRST EDITION

Designed by Sarah Lawson

Library of Congress Cataloging-in-Publication Data has been applied for.

ISBN 978-0-06-135160-0

09 10 11 12 13   ov/rrd   10 9 8 7 6 5 4 3 2

To Little Neil and Maggie

He dwelt in an isolated house,
because he was a leper.

—2 CHRONICLES

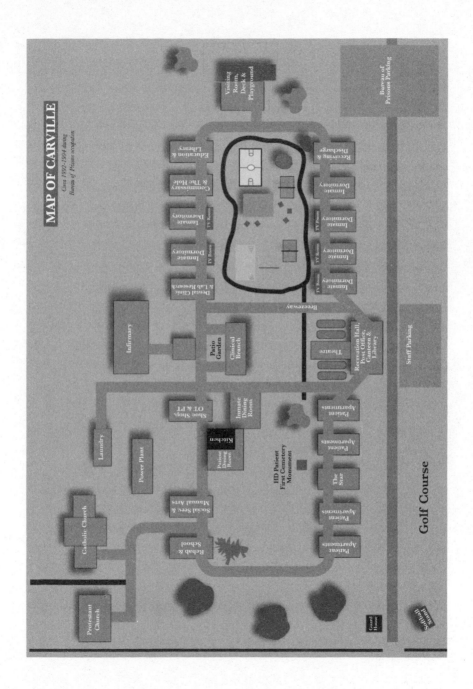

MAP OF CARVILLE

*Circa 1992–1994 during
Bureau of Prisons occupation*

Bureau of Prisons Parking

Visiting Room, Deck & Playground

Education & Library

Commissary & The Hole

Inmate Dormitory

TV Room

Inmate Dormitory

TV Room

Dental Clinic & Lab Research

Receiving & Discharge

Inmate Dormitory

TV Room

Inmate Dormitory

TV Room

Inmate Dormitory

TV Room

Breezeway

Infirmary

Patio Garden

Clinical Branch

Recreation Hall, Post Office, Canteen & Library

Theatre

Staff Parking

Laundry

Power Plant

OT & PT

Shoe Shop

Inmate Dining Room

Kitchen

Patient Dining Room

Patient Apartments

Patient Apartments

Social Serv. & Manual Arts

Rehab & School

HD Patient First Cemetery Monument

The Star

Patient Apartments

Patient Apartments

Catholic Church

Protestant Church

Guard House

Guard House

Golf Course

AUTHOR'S NOTE

For more than a century, Carville, Louisiana, served as the United States' national leprosarium. Individuals who contracted the disease were forcibly quarantined at its remote location on a bend in the Mississippi River. By the 1990s, the number of patients at Carville had dwindled to 130, the very last people in the continental United States confined because of the disease. The facility had hundreds of empty beds, so the Bureau of Prisons transferred federal convicts to Carville. *In the Sanctuary of Outcasts* is the story of the year I was incarcerated at the Federal Medical Center in Carville, Louisiana.

A Note on the Word *Leper*

I wish the word *leper* were not in our vocabulary. For the individuals who contract leprosy, this ancient term is deeply offensive: the label defines individuals solely on the basis of their disease and further alienates them from the world. Early in the book, I have included the term as I used it—in my own ignorance—when I first arrived at Carville. I lived with, watched, and ultimately forged friendships with the residents of Carville, many of whom welcomed convicts into their home. For this reason, I have used the term *leper* as sparingly as possible to depict the suffering caused by this branding, the misunderstandings about the disease, and the stigma associated with leprosy. My hope is that the book will reflect my gradual understanding of, and empathy for, this community of men and women who survived unimaginable injustice and tragedy. After the "Summer" section, as narrator, I do not use the word. For the remainder of the book, the term *leper* is confined to dialogue, sequestered within quotations.

PART I

My First Day
May 3, 1993

Live oak trees separate the front of the colony from the Mississippi River levee.

CHAPTER 1

Daddy is going to camp. That's what I told my children. A child psychologist suggested it. "Words like *prison* and *jail* conjure up dangerous images for children," she explained.

But it wasn't camp. It was prison.

"I'm Neil White," I said, introducing myself to the man in the guardhouse. I smiled. "Here to self-surrender."

The guard looked at his clipboard, then at my leather bag, then at his watch. "You're forty-five minutes early."

"Yes, sir," I said, standing tall, certain my punctuality would demonstrate that I was not your typical prisoner. The guard pointed to a concrete bench next to the guardhouse and told me to wait.

The grounds were orderly and beautiful. Ancient live oaks, their gnarled arms twisting without direction, lined the grove between the prison and the river levee. The compound—called "Carville" by the U.S. marshal who had assigned me to this prison—was a series of classic revival-style two-story buildings. The walls were thick concrete painted off-white, and each building was connected by a two-story enclosed walkway. Large arched windows covered by thick screens lined the walls. There were no bars on the windows. Nothing but screen between prison and freedom.

Through the windows I saw a man limping in the hallway. He stopped at the last arched window, the one closest to the guardhouse, and looked out. He was a small black man wearing a gentleman's hat. Through the screen his face looked almost flat. He stood at the window and nodded as if he had been expecting me, so I waved. He waved back, but something was wrong with his hand. He had no fingers.

I stood and stepped over to the guardhouse. "Is that an inmate?" I asked the guard with the clipboard, motioning toward the man behind the screen.

"Patient," the guard said.

"A sick inmate?"

"You'll find out," he said, and went back to his clipboard.

I looked back for the man with no fingers, but he was no longer at the window. I wondered if he had lost his fingers making license plates or in some kind of prison-industry accident. Or God forbid, in a knife fight. I returned to my bench wondering why he was roaming about instead of locked in a cell.

The prison sat at the end of a narrow peninsula formed by a bend in the Mississippi River, twenty miles south of Baton Rouge. The strip of land was isolated, surrounded by water on three sides. My wife, Linda, and I had driven ninety quiet, tense minutes north from New Orleans. We left the radio off, but neither of us knew what to say. As we passed through the tiny town of Carville, Louisiana, a road sign warned: PAVEMENT ENDS TWO MILES. Just outside the prison gate, I'd stood at the passenger window. Linda looked straight ahead gripping the steering wheel with both hands. I'd leaned in through the window to kiss her good-bye. A cold, short kiss. Then I watched her drive away down River Road until she disappeared around the bend.

As I sat on the bench, waiting for the guard, I resolved again to keep the promises I made to Linda and our children—that I would emerge the same husband, the same father; that I would turn this year into something positive; that I would come out with my talents intact; that I would have a plan for our future.

A guard in a gray uniform drove toward me in a golf cart. He stopped in front of the bench and stepped out of the cart. A tall, muscular black man, he must have stood six feet, four inches. A long silver key chain rattled when he walked.

"I'm Kahn," he said.

I introduced myself and held out my hand. He looked at it and said, "I know who you are."

I put my hand back by my side.

He picked up my British Khaki bag. It was a gift from Linda and a reminder of better times. I had packed shorts and T-shirts, tennis shoes, socks, an alarm clock, five books, a racquetball racket, and assorted toiletries, as if I were actually going to camp. Kahn tossed the bag in the cart and told me to get in.

We drove down a long concrete road that ran along the right side of the prison adjacent to a small golf course, and I wondered if inmates were allowed to play. We passed at least ten identical buildings that looked like dormitories. The two-story enclosed hallways that connected each building formed a wall surrounding the prison. The place was enormous. Enough room for thousands, I guessed.

I had done my research on prisons. Not as an adult, but in high school. I had been captain of my debate team. I understood the pros and cons of capital punishment, mandatory minimum sentencing, drug decriminalization, bail reform, and community-service sentences. I won the state debate championship advocating drug trials on convicts. I argued with great passion that testing new medications on federal prisoners would expedite the FDA's seven-year process to prove drug safety and efficacy, that the financial drain on taxpayers would be greatly reduced, and that these tests would give inmates an opportunity to earn money, pay restitution, and seek redemption, while thousands of innocent lives would be saved. When I was debating the merits of drug testing on prisoners, I never dreamed that I might someday be one.

Kahn stopped the golf cart at the last of the white buildings. He grabbed my bag as if it were his own now, and we entered through a metal door. The walls were newly painted, and the floor was well polished and shone like Kahn's shaved head. I walked behind him down a narrow hallway, and he pulled the chain from his pocket. He unlocked a door marked R & D. My heart skipped, and I felt panic coming on as we stepped inside.

Except for a wooden table, the room was empty. Kahn threw my leather bag on the concrete floor, positioned himself behind the table, and assumed a military stance.

"Front and center!" he commanded. I wasn't sure where to move. He put his hands on the table, leaned toward me, and yelled, "I said *front* and *center!*"

I stepped between the table and the wall and stood still facing him. I didn't want him to have to repeat himself again.

"Strip down," he said. I removed my shirt, pants, and shoes and took off my watch. I lifted each foot and pulled off my socks. "All of it," Kahn said.

I removed my underwear and dropped it on the floor. The concrete was cold on my feet. I held my hands at my side, but I wanted to cover my front. Once an athlete, I now sagged. My ritual of rich business lunches—seafood appetizers, fettuccini Alfredo, filet mignon with béarnaise, and chocolate decadence—coupled with an abundance of red wine at night had added forty pounds. Kahn rattled off a set of commands. *Lift your left arm. Now, your right. Bend forward, run your fingers through your hair.* After each command, Kahn paused and examined the exposed area. He continued. *Lift your penis. Lift your scrotum. Turn around. Face the wall. Lift your left foot. Now, your right. Bend over. Spread your cheeks.*

I glanced over my shoulder to make certain I had heard correctly.

"Bend over," he repeated, irritated, "and spread your cheeks."

I bent over and placed my hands on each side of my buttocks. I slowly pulled them apart. As I held my position, I felt blood rush to my face. I felt humiliated. I looked at Kahn through my legs. "You know," I said, "I won the DAR Citizenship Award in high school."

Kahn remained expressionless. I had hoped to disarm him, make him laugh so he would see I was not like the other men here, but he wasn't interested in my attempts at humor. He finally turned away and tossed a green shirt and a pair of green pants on the floor. I straightened myself up and examined my new uniform. The pants were too small, and the shirt was horribly wrinkled.

The suits and shirts I had worn on the outside were always professionally pressed. A perfect outward appearance, I believed, would accurately reflect the quality of my work and assure clients that my attention to detail had no boundaries.

"Do you have an iron?" I asked Kahn as I held up the shirt, examining its poor condition and missing buttons.

Kahn didn't answer. He didn't even look at me. Instead he opened my bag, turned it upside down, and emptied the contents onto the table. He quickly sorted through my belongings, tossing to the floor items he said would be sent home. The items he kept on the table would stay with me. He held the stack of books I'd packed and told me to pick two.

I'd brought along some southern classics I had never made time to read—*A Confederacy of Dunces* by John Kennedy Toole, *Good Old Boy* by Willie Morris, and *The Moviegoer* by Walker Percy—but two other books were more important to me.

Every year, beginning with my eighth birthday, I could count on one Christmas gift from my father: a copy of *The Guinness Book of World Records*. From the world's fastest human to the tallest radio tower, from the wealthiest family to the largest blue whale, from the most consecutive jumps on a pogo stick to the world's biggest pancake, the people in the pages of the Guinness book were not ordinary. These individuals had status and prominence and immortality. They were one of a kind.

More than anything else, I'd wanted to be listed in the book. But I had a few early setbacks. My pogo stick got stuck at jump 2,009 when the oil burned off the pole. My growth spurt hadn't taken off like Robert Wadlow's, so I had to face the fact that I might not grow to surpass eight feet, eleven inches. And while practicing to break the world-record discus throw—a record I just knew was within my reach—I sent a two-pound weight through the back windshield of my mother's car.

For me, everything was a race. I raced against a clock when I mowed our lawn or held my breath underwater or scarfed down food. I was completely unconcerned with the *exact* record I would break, as long as I ultimately accomplished one act, one conquest that no other human had ever achieved.

I told Kahn I would keep *The Guinness Book of World Records*. I chose the Bible as my second book because I had hidden photographs of Little Neil and Maggie in the back.

With a metal-tipped vibrating device, Kahn etched something on the back of my wristwatch—a Christmas gift from Linda and the kids. Kahn tossed the watch back to me. "03290-043"—my inmate number—was scratched on the back.

"Do you have any money?" he asked.

I had a $20 bill in the side pocket of my bag. "Paper money is contraband. Inmates can only have coins," he explained and handed me two rolls of quarters. He started toward the door. I couldn't let him go without asking about the sign on the door. "Research and Development?"

Kahn looked confused and annoyed. "Receiving and Discharge," he answered.

"But I saw a patient earlier," I said. "What kind of—"

"Hansen's disease," Kahn interrupted, walking toward the door. Without looking back, he added, "It used to be called leprosy."

Kahn left the room and locked the door behind him.

CHAPTER 2

Leprosy. Kahn had to be wrong. Surely, healthy people—even inmates—would not be imprisoned with lepers. But that *would* explain the man with no fingers. Everybody knew lepers' body parts fell off. Or maybe Kahn was just beginning the mind games I'd seen guards use in movies to break prisoners.

A nurse rushed into the room and opened two folding chairs. She told me to sit at the table, and she asked me a series of routine questions about drug use, smoking, chronic diseases, and depression, to which I answered no.

"Any family history of mental illness?" she asked.

I hesitated. "Define mental illness."

The nurse suggested I tell her about any family members who might fall into that category, so I explained about my great-aunt, who bought seventy pairs of shoes in a single day before she was committed, and about my grandmother, who did a couple of stints in the state mental hospital and then ran for president, twice—

The nurse interrupted. "Of the United States?"

"Yes," I told her, "but it all happened when she was off her medication." Then I mentioned that my mother sees auras and claims to have been Mary, Queen of Scots in a former life.

She interrupted again. "I'm going to mark this yes."

When she finished, I asked if there were really lepers living here.

"They prefer to be called Hansen's disease patients," she said. "But, yes, about 130 live here."

I asked if they were contagious and if we ever got close to them and, if so, was there some way to get transferred to another prison. The nurse cut me off and said I'd hear all about it at admission and orientation.

My mind raced as she collected her paperwork. I could recover from a year in prison, but I couldn't put my life back together with a missing hand or a deformed face. That would be like a life sentence. If I caught leprosy, I would lose my family, never be able to get close to Neil and Maggie. I was frantic, but I had no way of letting anyone know what was happening to me. I was completely helpless.

I gathered the two books and the few clothes Kahn let me keep. Then the nurse escorted me to a hallway and gave me directions to my room. It seemed strange that I'd been left to wander around without a guard or escort.

The hallway smelled like my grandmother Richie's farmhouse, that earthy scent of dust in a closet that had been closed up for years.

Arched windows lined the elevated hallway that went on as far I could see. The sunlight, tinted by thick screens, threw bands of symmetrical amber light against the wall, like dozens of sepia tombstones waiting to be engraved.

The hallways formed a quadrangle, and inside was a lush, almost tropical, courtyard with banana trees and mimosas, oaks and azaleas. It was not at all what I imagined a prison would be like. It felt tranquil, like a beautiful island paradise I'd expect to find in Hawaii.

Through the screen I saw inmates shooting basketballs, tossing horseshoes, and walking around a concrete track. I heard soft chatter and laughing and the sharp snap of dominoes hitting a table.

I walked out into the courtyard. The crowns of the buildings were contoured with extravagant sculptured designs and plaster cornucopia scrolls. The pungent smell of fresh-cut grass reminded me of the slow summer days of my childhood.

Men were lounging in the shade. Some were dressed in khaki uniforms, others in green like mine. Some of the men were old; others looked not much past their teens. A few were in wheelchairs. There were blacks, whites, and Hispanics. At one table sat four of the most obese men I'd ever seen in person. They were playing dominoes. They didn't look much smaller than Robert Earl Hughes, the world's fattest man, whose photo I had studied at night in *The Guinness Book of World Records*.

Three men were sunbathing on a shuffleboard court. Another man was zipping around the grass on a small, motorized four-wheeler pulling a trailer full of garbage bags. He drove the vehicle in my direction and stopped in front of me. He turned off the engine and let out a loud howl like a coyote.

"You know they got lepers here, don't you?" he said.

"I've heard."

"And you're a convict, right?" he asked.

"I guess so."

The man smiled and said, "Then that makes you a lepercon!" He laughed, threw his head back, and howled again. Then he cranked the engine and drove up a ramp and into a hallway.

I noticed a few inmates looking my way, so I hurried up the ramp and into the hallway that led to my dorm.

As I walked down the corridor, holding everything I now owned in my arms, an elderly black woman in an antique, hand-cranked wheelchair rolled toward me. Two long, vertical chains ran from the handles to each wheel. She cranked the wooden handles like a child pushed pedals on a bike. With each crank, the wheels on her chair turned. The skin on her hands was shiny and cracked. She wore a turquoise striped dress that hung from the seat of her chair like a wrinkled curtain. She had no legs.

With each churn of the handle, with each rotation of her hand, the wheelchair moved closer to me. The woman's eyes were bright in contrast to her dark sockets. Her hair was silver and black. Her fingers gripped tightly around the wooden handles. With each crank the wheelchair wavered. Her earrings swayed with the tempo.

This was a prison for men, which meant she wasn't an inmate. And she certainly wasn't a nurse or a guard. I made eye contact and smiled like I might have to a beggar in the French Quarter. I took satisfaction in being polite to the down-and-out. Had I encountered this woman on the street, I might have stuffed a few bills in her cup, but here, I was wary of getting too close. She smiled and looked me directly in the eye. I stepped to the side of the walkway to make room for her to pass, took in as much air as possible, and held my breath. I had perfected a technique in elementary school when my teacher, Ms. Cauthen, who had terrible halitosis, would lean over my desk. I would hold my breath and put on a tight-lipped smile. When she moved away from my desk, I would cover my mouth with my shirt-sleeve to filter the air and escape from the particles I imagined she had left behind.

I held my breath and smiled at the old woman, hoping to mask my apprehension. Slowly, she cranked her way down the corridor. Passing me she chanted, "There's no place like home." Her voice was worn-out, but sweet. I stood perfectly still and said nothing. Once she passed and could no longer see me, I put my belongings on the floor, covered my mouth with my shirtsleeve, and exhaled. I stood in the hallway with my mouth covered. She chanted again, "There's no place like home." I watched her slowly roll away and disappear around the corner.

CHAPTER 3

My building was called Dutchtown, named for a neighboring community on the Mississippi River. Inside, Dutchtown looked and smelled like my freshman dormitory at Ole Miss. The hallway floors were shiny and polished and eight doorways lined each side. The prison cells were really just rooms with linoleum floors and stucco walls, but all the doors had been removed. The frames still had screw holes from the hinges. As I walked to my room, I saw a man in a bathrobe and flip-flops walk across the hall and into another inmate's room. Obviously, I wasn't going to have much privacy.

Upstairs, I found number 204, my assigned room, and peered through the entrance. There were three small cots, a desk, a chair, metal lockers, a small closet, and a sink with a mirror over it. I didn't expect to have a mirror, but I was glad to see it. If I had to wear wrinkled clothing, at least I could keep well groomed and take some pride in my appearance.

In one of the beds, a man lay on his back reading. An open magazine entitled *Cutis*, a medical journal of some sort, was propped upright on his chest, hiding his face. I stepped into the room, and the man let the magazine fall. He took off his reading glasses and squinted to get a good look at me. He looked distinguished, even in prison attire. He was in his midfifties, with a slightly receding hairline and a neatly trimmed salt-and-pepper beard. He stood up, held out his arms, and said, "Thank God, you're white." He introduced himself as Victor Dombrowsky. "But everybody calls me Doc," he added. "Welcome to Carville," he said, "where they quarantine lepers, where the petrochemical companies discharge their waste, and where they send the likes of us. You're now officially part of a human garbage dump."

Doc picked up an armload of magazines from the cot next to the window, explaining that he'd been using my bed as a desk.

I placed my shirts, shorts, books, and quarters in an empty locker, and Doc went back to reading his medical journal. As he read, he asked about my family, my hometown, the length of my sentence, and my crime. I could see Doc's profile as he read, but he never looked away from his magazine.

"Did you testify against anybody?" he asked.

"No," I said, moving to make up my bed.

"Good," he said after a beat, "'cause I hate fuckin' rats."

I put the coarse, government-issued sheets and gray wool blanket on my bed. I would miss the Egyptian cotton sheets that felt so good at home.

As I struggled to put my pillow into its case, Doc said, "Tell them you've got neck trouble. They'll give you an extra pillow."

I mentioned to Doc that I had been in the magazine business. He put his journal down, sat up, and showed a touch of enthusiasm. "Maybe you can help me," he said. "I've invented a device, and I need a good marketing man."

I asked him about the invention. Doc hesitated. "Well . . . it's an injection device. When it's loaded with a certain drug combination, it cures . . . impotence."

"Where do you inject?" I asked.

"That's my major hurdle," Doc said, like he'd been over all this a thousand times before. "There's this stigma about giving yourself an injection in the base of the penis." He paused and took a deep breath. "It doesn't hurt. I've done it. With clever marketing, I can get around it, don't you think?"

I stared at the man who would sleep a few feet away from me and tried to shake the mental image of him injecting himself in the penis. I wanted to tell him that I thought he was insane and no amount of marketing would ever overcome the horror of his invention. But I told him I probably needed to learn more about his product before I could offer an opinion. The last thing I wanted to discuss on my first day of prison was erectile dysfunction.

"Yeah, hell," he said. "It's your first day. I'll have plenty of time to fill you in later on." He started reading again. "I've got a prototype," he said, "but I can't find anybody on the outside who will test it."

I wanted to change the subject. "What are the other inmates like?"

"Idiots," he said, "complete, total idiots."

Doc said more than four hundred inmates were serving time in Carville. About half of them suffered from some health disorder like heart disease, cancer, or AIDS. Some were in wheelchairs from gunshot wounds, degenerative disease, amputations, or congenital deformities. Since Carville had a hospital, most federal inmates with serious health problems were sent here. But Doc said Carville also had drug dealers, Mafia guys, even murderers. I thought Carville was supposed be a minimum-security prison, and Doc's description of the other men was unnerving, especially since our room had no doors.

I opened the small closet door looking for a place to hang my pants, but the floor and every shelf were filled with stacks of medical journals on dermatology, clinical oncology, metabolism, and other medical specialties.

"I'll make some room in there if you need to hang up some clothes," Doc said.

"You subscribe to all these?" I asked.

"They're free," he said.

Pharmaceutical companies sent complimentary subscriptions to any physician who requested copies. After Doc's conviction, his medical license was suspended in every state except Tennessee, where he'd attended medical school. Since the subscription cards came from a federal medical center, the publications assumed Doc was practicing medicine. He received more than sixty journals a month.

"You read all of them?"

"I may be the only doc in America who actually has the time to read them all," he said, sitting up on his bed. "I've learned as much about medicine in here as I did in med school."

I figured Doc must know about leprosy so I asked about the patients.

"Grotesque, aren't they?" Doc said. "They used to live in this very room."

"Is it contagious?" I asked.

"Not supposed to be," he said, pausing, "*if* they take their medication. Nobody really knows how the disease is spread." He added that the mystery surrounding it made him nervous. Doc explained there were a number of theories about how leprosy is contracted, including intimate skin-to-skin contact or eating an infected armadillo. "The most likely theory is inhalation of an infected droplet," he said with a shrug. "Who knows?"

I wasn't worried about eating armadillo, or even skin-to-skin contact. Those, I could avoid. But if breathing in a droplet could cause infection, a sneeze or a cough might be enough.

This just kept getting worse. I could be outside doing something worthwhile—paying creditors, taking care of my family. I would gladly have accepted a fine or home confinement or work release or all of them combined, but I didn't deserve this. Not a leper colony.

Just then Kahn stepped into the room. "You got a job assignment," he said. "Follow me."

"Where?" I asked.

Kahn snapped back, "You don't get to ask questions anymore."

CHAPTER 4

The walk took about five minutes. I followed Kahn, winding through the old corridors. After Doc's remarks about leprosy, I was careful not to touch anything. We passed a manicured garden enclosed in a small courtyard. Tiny shrubs and patches of yellow and blue flowers bordered brick paths. It looked like a place where an English family might gather for tea. We turned a corner, and I caught a glimpse of four or five nuns as they hurried into one of the buildings. Through a corridor window, I saw a small monk riding a bicycle through a pecan grove. This place was bizarre, like something out of *Alice in Wonderland* or *The Twilight Zone*. Nuns and monks. A leper with no fingers. A man who howls like a dog. A doctor with an impotence injection device. Inmates fat enough to be in a carnival. A guard who squelches my questions, but seems just fine with prisoners sunbathing. And a legless woman chanting like Dorothy in Oz. How the hell did I end up here?

It turned out Kahn was escorting me to the cafeteria. A hand-carved wooden sign, the kind you might see at a southern diner, was nailed over the door. It read MAGNOLIA ROOM. Kahn left me at the door and told me to go inside to get my assignment.

In the cafeteria office a guard asked, "You got good handwriting?"

Actually, my handwriting was superb. I had spent years perfecting it. I had even invented several of my own fonts. In high school I was always asked to help create school signage, and I learned early on that teachers were reluctant to mark up a beautifully written assignment. I loved the praise that came along with fastidious manuscripts. The notes teachers wrote—*Beautifully written. A pleasure to read. If only all my papers were so neat!*—inspired me to strive for even greater perfection.

"Penmanship is one of my strong suits," I told the guard.

He handed me a menu for the day's meals and a set of dry erase markers.

"Write the menu on the board in the patient cafeteria," he said. "Right through there," he added, pointing at a door.

As I walked through the kitchen, I saw an inmate stirring a huge pot of soup. He noticed the markers and yelled, "Write big 'cause them lepers can't see worth a shit!"

I would think twice before bragging again. I pushed through a heavy swinging door and saw them. Fifty, maybe sixty leprosy patients sitting at the tables. I scanned the room for other inmates, but I was the only one. The menu board was on a wall on the opposite side of the room. I kept my eyes on the floor as I moved through a maze of wheelchairs and walkers, canes and crutches.

An Asian man sat at a table in the middle of the cafeteria. He stared at me as I approached. A white growth completely covered one of his eyes. Between his two digitless hands, he balanced a pork chop. His good eye followed me as I walked. The skin around his mouth and chin was covered with a dark blue ointment, and his hands were shaped like mittens. I didn't want to stare, but I couldn't help myself. The pork chop slipped from his grip and fell onto his plate. He mumbled something, his mouth full of chewed meat.

I passed men and women with odd-shaped noses, discolored faces and disfigured hands, oversized sunglasses, irregular-shaped shoes, and stumps from missing limbs. I held my breath, hurried toward the menu board, and stood as close to it as possible, my back to the lepers.

I couldn't believe lepers still lived in America. Leprosy was something that happened in third-world countries. I had always imagined lepers—dangerous and grotesque—the way they were described in the Bible or portrayed in Hollywood films, being forced out of cities and told to wear bells or clappers to warn travelers of danger.

I didn't want to breathe the air, or accidentally brush up against one of them, or get close enough that the infection could reach out, take hold in my body, and turn me into a horror.

Focus on the menu, I told myself.

I took off my apron and wiped away the sloppy handwriting that described yesterday's meals. I started to copy the menu from the paper the guard had given me. Then I felt a tap on my back.

I froze. I didn't turn around. I didn't want to face these people. Doc was wary of the disease, and I figured he knew plenty.

I felt the tap again. I reluctantly turned and saw the man with one white eye. He pushed himself up out of his wheelchair.

"Use the *purple!*" he yelled. A drop of spit flew from his mouth. It hit my cheek. I took a half step back.

"We can see *purple* the best!" This time the spit landed on my apron.

I nodded. I desperately wanted him to go away. He sat down in his wheelchair and rolled toward the exit. I scrambled to find a clean corner on my apron to wipe my face. My heart pounded, and I felt dizzy. I couldn't believe he had just spit on me. Then I remembered Doc's words: *inhalation of an infected droplet.* I took shallow breaths. A cold burn ran through my cheeks. Part embarrassment and part rage, but mostly shock.

I needed to pull myself together. I needed to stay calm and be reasonable. Right now, I had a job to complete, and I wanted to finish as soon as possible.

The room had an overpowering sweet smell like the syrupy stench of fresh-cut banana trees. Combined with the thick, greasy odor of fried pork chops, the smell made me nauseated. I gathered myself and wrote in big purple letters: *French Toast. Tuna Casserole. Meat Loaf.* The side items—*Mashed Potatoes, Green Beans, Tater Tots*—I wrote in yellow.

As I carefully transcribed the menu, most of the leprosy patients exited the cafeteria. Some used walkers, but most were in wheelchairs. Thankfully, they left me alone. I reboxed the markers. Then I saw the old woman in the antique wheelchair, the only one left in the room.

She cranked her wheelchair toward me. She stopped a few feet away, not too close, and uttered the same odd incantation. "There's no place like home." Aware, I think, of my discomfort, she looked at me and said, "Hope you get back soon, 'cause there's no place like home." She smiled and cranked her wheelchair out of the cafeteria. When she reached the exit, she called out again, "There's no place like home."

An inmate who had come in to mop the floor whispered to me. "That lady," he said, pointing toward the old woman, "she got the leprosy when she was twelve years old. Her daddy dropped her off one day and never came back." Then he asked, "Still feeling sorry for yourself?"

I guessed the woman was close to eighty. That would mean she'd been here for about sixty-eight years. I was going on my sixth hour.

(left) *Maggie's ballet recital.*
(below) *Little Neil at
Walloon Lake. Photographs
I attached to my locker with
toothpaste.*

CHAPTER 5

Back in my prison room, I wrote a letter to Linda and the kids. I had promised to write every day. I had also promised Linda I would be truthful about everything. But I couldn't tell her I was living in a leper colony and that one of them had spit on me, or that my cellmate wanted me to help market a penile injection device.

I was a master of positive spin, but this time I was stuck. I had no good words to write. I put my pen down. I rubbed the spot on my cheek where the leper had spit on me.

"Doc?" I asked. "Do body parts really fall off?"

"That's a myth," he said. "They lose hands, feet, even legs, but usually by amputation." Doc walked over to the mirror, wrapped a washcloth around the head of a disposable razor, snapped it in half, and carefully removed the blade. As he unbuttoned his khaki shirt, he explained that leprosy was a neurological disorder. In some cases, if an infection goes untreated, the body absorbs fingers and toes. "It's just awful," he said. Doc took his shirt off. He picked up the blade from the shattered disposable razor, turned his back to the mirror, looked over his shoulder, and began, very carefully, to carve tiny incisions in the skin around a mole on his back.

While I stared, Doc continued to list the symptoms of leprosy—skin lesions, blindness, erosion of the nose, diminished immunity, and huge nodules on the face and hands. Some of the victims, he explained, looked perfectly normal. Others had minor deformities like a misshapen hand or foot. Still others had discolored skin, or feet so swollen that shoes wouldn't fit, or faces that appeared to have been dipped in acid.

Doc said the disease eats away at the extremities—cooler parts of the body. If leprosy is left untreated, the victim's body literally begins to disappear. And there are no tests to predict who might be susceptible. Doc put the razor against his skin again and winced. I had watched as long as I could.

"What are you doing?"

"Removing a spot," he said, gesturing with the razor in his hand. "Could be cancerous."

"Aren't there doctors here?"

"Idiots." He scowled. "They'll kill you." He made another cut. A trickle of blood ran down his back. He pressed a tissue against the wound and sat on the bed again. "The worst thing about leprosy," he said, "is the nerve damage. Total loss of sensation. The poor bastards can't feel it when they hurt themselves."

That night as I lay on my bed and thought about the old woman with no legs, being left here as a child, never to see her family again, it made me think about my children. I pulled their photographs from the back of my Bible and spread them on my bed. I'd brought my favorite shots. Little Neil, his strawberry hair almost blond from the sun at Walloon Lake, Michigan, where we vacationed in the summer. Maggie, so tiny, in her pink ballet outfit during a recital at the Louise McGehee School in New Orleans's Garden District. One photograph of the four of us after an Easter church service sitting in the courtyard garden at St. Peter's by-the-Sea. If I taped the prints on the inside of my locker door and left it open at night, I would be able see the photographs from my bed.

"Doc," I asked, "can I borrow some tape?"

"Tape is contraband," he said. "Use toothpaste."

"Does that work?"

Doc grabbed a tube from a box under his bed and tossed it to me, "Tooth-*paste*," he said.

I turned the photographs facedown on my bed and carefully spread Colgate on the back of one of the prints. I pressed it against the inside of the locker, and it stayed. I carefully applied toothpaste to the other photos and attached them to my locker door. Then I noticed my two rolls of quarters were gone.

"Damn it!" I said. "Somebody took my money!"

Doc chuckled. "Rumor has it there *are* criminals here."

PART II

Summer

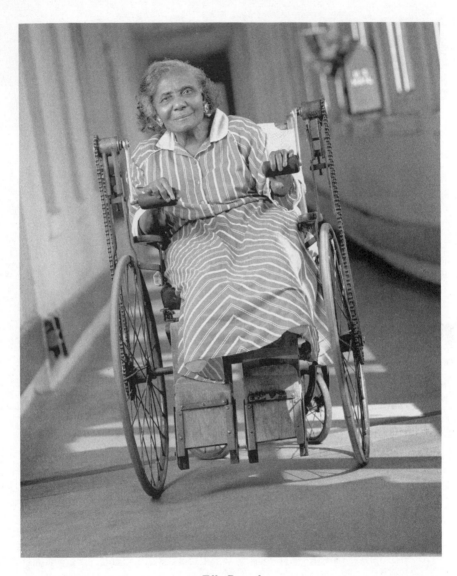

Ella Bounds

CHAPTER 6

The guard banged his flashlight against the end of my cot and pointed it at my face. I sat up, covered my eyes, and told him I'd be right down. It was still dark outside. I looked over at Doc's alarm clock. 3:45 A.M.

Doc moaned, "You've gotta get another job."

I had been assigned to the 4:00 A.M. to 1:00 P.M. shift in the cafeteria, six days a week. My pay: fourteen cents an hour. I was short on sleep. The inmates on my hallway stayed up until midnight playing dominoes, and instead of placing the dominoes down on the table, they slapped them down. The sound was like a firecracker. When the guards finally did break up the games, the snoring started. A half-dozen men, most of them over three hundred pounds, suffered from sleep apnea. They wore breathing machines, and their snores echoed up and down the hall. If the noise weren't enough, two lights in the hallway remained on twenty-four hours a day. One of them, outside our doorway, cast a bright beam of light onto my bunk. I learned to sleep with my arm over my eyes.

I dressed in one of my green uniforms and walked the long, empty corridor toward the entrance where the leprosy patients lived. The hallways smelled musty and sweet. The door dividing the two sides was secured at midnight each night and was still locked, so I walked downstairs and joined a group of other inmates who were gathered in the prison courtyard waiting to be escorted to the cafeteria.

The buildings and corridors of the colony formed two huge quad-rangles. The one closest to the river was reserved for the leprosy patients. The patient quadrangle contained a few gardens planted by the patients and a tombstone commemorating the first one hundred individuals who died in Carville, some of whom were identified by nothing more than initials. The other quadrangle was ours. Surrounded by the two-story concrete walkway, the inmate courtyard was outlined by a quarter-mile serpentine walking track. Inside the track, the prisoners had access to a weight-lifting area and stationary bikes. The cafeteria building had been built in one corner of the leprosy patient quadrangle.

The stagnant summer heat, even at this hour, was heavy. The colony rarely felt a breeze because it sat at the base of the levee. The dead air and humidity pushed temperatures to a hundred degrees on most summer days. Within minutes of venturing outside, I felt perspiration soak into my shirt.

As we waited for the guard, I noticed the other inmates wore thick gloves, winter caps, and heavy jackets. I asked Jefferson, a skinny kid from New Orleans, why they wore cold-weather gear.

"He don't know shit, do he?" he said to the others. Then they tapped one another's fists. I was out of my element and felt stupid. A guard finally arrived, and we walked together to the cafeteria.

I had two jobs: washing dishes and writing menu boards. It seemed strange that the guards would bother with a menu board since we didn't have a choice about what was served. But food was important here. Everything I had ever read about prison suggested that weapons or drugs or whiskey or some other contraband would be the convicts' primary concern. Here food was currency, particularly fruit, which was reserved for the leprosy patients. Lemons, bananas, and oranges brought up to five dollars each. Strawberries, cantaloupes, and honeydew melons were rare delicacies. The only way for prisoners to get fruit was for an inmate to smuggle it from the cafeteria.

The lunch menu included barbecue pork, so on the menu board, I sketched an illustration of President Clinton skewering a pig. I finished both boards in less than an hour. Since I didn't have to be in the dish room for another two hours, I walked through the industrial kitchen to see what the other inmates were doing.

The kitchen was empty. I checked the dish room and the dry goods warehouse. The stoves and cutting stations and mixers sat unused. I opened the door to the large walk-in cooler. The shelves were stacked with vats of mayonnaise, large blocks of butter, gallons and gallons of milk, and boxes of lettuce, tomatoes, and other vegetables. At the rear of the cooler, I saw a boot propped on top of a produce box. Jefferson and five other inmates sat in the back of the cooler, winter caps pulled tight over their ears, hands stuffed into coat pockets. The men were wedged between some half-empty boxes stacked against the back wall, sound asleep. With each exhale, soft steam floated out from their noses and mouths.

I went back into the main cafeteria and made a cup of coffee. Dark, strong New Orleans chicory coffee. I picked up a copy of *USA Today*. The room was quiet and still, and the smell of buttery dough reminded me of my high school cafeteria. I sat, sipped my coffee, read the paper, and wondered how breakfast would ever be ready with everyone napping in the cooler. The "Life" section of the paper featured a soon-to-be-released summer blockbuster film. I thought about taking Little Neil to see it. We would go on opening night, maybe to the Prytania Theatre, where the lines wouldn't be as long, or maybe Canal Center. We would have a quick dinner at Gautreaux's, get to the cinema early, and grab a seat in the front row with the other kids.

For a moment, lost in the story, I had forgotten. My son would see the movie, but not with me.

Through a lattice wall dividing the prisoner and patient dining rooms, I saw the old woman in the antique wheelchair cranking around in the leprosy side of the cafeteria. She saw me, too, and motioned for me to come over. I was growing accustomed to her chanting and her smile. And even though at first I didn't want to breathe the same air, for some reason, she seemed harmless and sweet. Maybe it was learning about how she had been abandoned as a young girl.

I folded the paper, grabbed my coffee, and walked around the lattice.

"You're up early, aren't you?" I asked.

"Yep," she said. "Already had my bath, too."

I asked if I could bring her some coffee.

She nodded.

"How do you like it?"

"Black with sweets-and-low," she said. "Lots of sweets-and-low."

Rather than handing her the coffee cup, I put it on the table. I wanted to talk, but I also didn't want to get too close. I sat down opposite her. Each square table was set with two chairs—the other two sides were left open for wheelchairs. She picked up the plastic coffee mug with both hands. The skin on her fingers looked like she had just applied lotion. She had all ten of her fingers. No sign of her body's absorbing digits like Doc had described, but she was mindful with the coffee. She kept her eyes focused on the mug because, I assumed, she could not actually feel it. She took a sip and carefully placed the cup back on the table.

"My name is Neil," I said, hoping for her name in return, but she just smiled and nodded. "What's your name?"

She answered, but I couldn't understand her. I wasn't sure if she said Cella or Ella or maybe even Lola. I asked her again.

She spelled it out in a gravelly voice, "E-L-L-A."

We were all alone and I had an hour before the inmates arrived for breakfast, so I continued. "I live in New Orleans," I said, "but Mississippi is home. What about you?"

"I was born in Abita Springs, Louisiana," she said. "But this is my home." She picked her coffee up again. She held the cup between her palms and shifted in her chair. "How long you gotta stay?"

"About a year."

"Long time," she said. "Long time."

We sat quietly for a moment, then Ella let out a soft sigh. "You can be my guest," she said, "least 'til you get back to your place."

Ella was trying to make *me* feel better, even though she had been here for decades.

"So," I asked, "how did you end up here?"

Ella leaned back in her wheelchair, settling in. "Abita Springs," she said in a whisper. "Nineteen hundred and twenty-six. I was in grade school."

According to Ella, a doctor had visited the one-room schoolhouse to administer shots. The raised oval spots on her leg where the pigmentation had disappeared had caught his attention. He pricked the blotches with a needle. Ella felt nothing.

"Next week, white man drives up," Ella said, "and I seen the Carroll boy pointin' outside. 'Oooh, Ella,' he say, 'bounty hunter fixin' to carry you away.' I look out and seen the man leanin' on his truck, wearin' the dark glasses, arms crossed all tight."

A hand-painted sign—large enough to be seen from neighboring farms and which would later be nailed to the side of her family's tenant house—extended from the back of the white man's pickup truck. Ella couldn't read the long word scrawled in large red letters. Later she would understand: "Quarantine."

The schoolteacher put a hand on Ella's shoulder, pulled her up from her desk, and led her outside. The other children ran over to the window. The teacher walked her across the small schoolyard toward the truck that idled at the edge of the field. The bounty hunter uncrossed his arms and pushed back his coat to expose a pistol. The teacher stopped and took her hand off Ella's shoulder. The man pointed to the back of the truck, and Ella climbed in.

As he drove away Ella looked out through the wooden slats. Her teacher stood with her hands over her mouth. Her classmates' frozen faces filled the schoolhouse window.

Ella sipped her coffee and took a break. I didn't say a word. I waited for her to go on with her story. For the first time in as long as I could remember, I was patient. I had no place to go. No meetings. No deadlines.

There was something remarkable about this woman. The way she held herself, and her eyes. She seemed to possess unwavering confidence. Or maybe it was strength. But at the same time, she was gentle and friendly.

For a moment, I had forgotten she might be contagious. She was so vibrant. My mother would have been able to tell me all about her wonderful aura. It was hard to believe Ella carried a debilitating disease.

A guard pushed through the swinging kitchen door and called out for me. "Inmate!" he yelled. "Get over here."

I hurried toward the kitchen. The guard was a short white man with red hair and a moustache that needed to be trimmed. His pants were too tight. The gray material was stretched tight against his thighs. He looked uncomfortable.

"No fraternizing with patients," he said.

Ella heard and looked directly at the guard. "We jes talkin'."

"Yes, ma'am, sorry to interrupt," he said, with deference. "I need this inmate in the kitchen."

Ella called out to me, "See you tomorrow."

CHAPTER 7

Toward the end of my first week of work in the cafeteria, I was drawing a hot dog in handcuffs on the menu board when a guard told me to report to the visiting room at 9:00 A.M. for admission and orientation.

About twenty other prisoners were already seated at round tables. On one side, a group of young black men slumped down in their chairs. On the other sat an assortment of white and Hispanic men, young and old, and a couple of men in wheelchairs. They all wore orange jumpsuits, the kind criminals wear during transport by the U.S. marshal. They looked as if they hadn't slept or bathed in days. I was the only inmate wearing green. I felt uneasy and took a seat in the back away from the others.

Three men sat at a long table in the front of the room: a prison guard, a man who was dressed like the surgeon general, and the monk I'd seen on my first day. The monk stood and held up his arms. The room quieted.

"Hello," he said in a soft voice, "I'm Father Reynolds. I'm glad you're all here." He realized what he'd said and tried to back up. "I don't mean I'm glad you're *here*. What I mean to say is—uh—as long as you have to be here . . . I'm glad to see you."

Father Reynolds stammered nervously for a few minutes. When he finally calmed down, he told us we were welcome at the Catholic church on Wednesday evenings or Sunday afternoons. He added that we could attend the Sunday service with our families during visiting hours. Then he prayed. I was relieved that I would be able to go to church with Linda and the kids when they visited. I had assumed we'd be without church together for a long time. I looked forward to holding Linda's hand while a priest talked about forgiveness.

The guard reviewed the prison rules and gave us each a set of written regulations. He covered inmate boundaries, visiting hours, contraband, and requests for medical needs. He spent an inordinate amount of time on mail restrictions, emphasizing that nude photographs of spouses were prohibited. He also warned that any mail containing pubic hairs would be confiscated and discarded. The guard cautioned that any violation of the rules would result in time added to our sentences. I tried to pay close attention, but I was distracted. One of the men in an orange suit, a black man in his late twenties, had turned his chair away from the presenters. He didn't listen to what the guard said; instead, he stared directly at me, squinting like he might need glasses. Every so often he shook his head and looked around the room like he expected others would be staring at me, too. He was a small man with huge teeth. He didn't look particularly dangerous, but I realized I probably wasn't good at gauging that sort of thing. I tried not to look in his direction, but I could feel his stare.

After the guard finished, a few inmates asked questions about money and television access. Another asked if the female guards were allowed to strip-search us. "You wish," the guard said.

Then the man who stared at me put his hand in the air and turned back toward the front. "I heard there was a 50 percent chance we was gonna be *leopards* when we get out of here." The other men in orange nodded and said they had heard the same thing. The man dressed like C. Everett Koop said he would cover that after a break, but he assured us we were not at risk.

We were released for fifteen minutes. Most of the inmates went outside behind the building to smoke. An elaborate wooden deck had been built onto the back of the visiting room. Picnic tables and bench seats were scattered around. The deck led to a small grassy yard surrounded by a low picket fence. A wooden playground set built to look like a pirate ship had been erected for the children of inmates. Neil and Maggie would like it.

I sat on a bench and listened to the inmate with big teeth announce to his friends that he hoped he *did* turn into a *leopard* because he could sue the prison for a million dollars and he would be the richest damn *leopard* in America. Then he noticed me sitting alone. He motioned to his friends and walked toward me with five of his orange-clad buddies in tow. "Goddamn!" he announced. "You look *just* like motherfuckin' Clark Kent!"

His friends laughed. I straightened my glasses.

"What you did?" he asked. "Fuck the judge's daughter!?" His voice, high pitched with the tempo of a comedian's, didn't sound dangerous at all. He talked loud and laughed at his own words. The other inmates moved closer.

I told him my crime was bank fraud.

"You a goddamn bank robber!?"

"No, no. Bank *fraud*." I explained I had encountered some difficulty with cash flow. My crime, I told him, involved the transfer of checks from one account to another in order to buy time to refinance my magazine business.

"I don't know nothing about no checks," he said, "but let me ask you a question." He looked around to make sure he still had an audience. "Did you take money from a bank you wasn't supposed to have?"

The other inmates waited for my answer. "Yes." I nodded.

"Then you're a goddamn bank robber!" I didn't argue the point. No one would have heard me over the hand slapping and laughter anyway. He leaned toward me. "How much you get?"

I told him I used the money for payroll, taxes, printing, and other publishing expenses. I could tell he didn't believe me.

"How much the bank lost?"

"Two banks were involved, actually," I said. "Together their losses were about $750,000."

He looked excited. "So, how much you got?"

"I don't have any of it," I reiterated. "I paid bills."

"I been in jails all over this country," he announced, still playing to the other inmates, "and you the *stupidest* damn criminal I ever met."

I was embarrassed to be the brunt of his routine, but he had a point. I was in prison and I had nothing to show for it. I had to admit the guy was pretty entertaining.

"By the way," I said, "my name is Neil."

"Ain't no more," he said. "You got a prison name now. And it's motherfuckin' Clark Kent."

I asked his name.

"They call me Link."

"Why Link?"

Another inmate interjected, "As in 'the missing . . .'"

I stood and held out my hand, "Nice to meet you, Link."

"Goddamn!" he said, looking at my extended hand. "This is prison. You ain't got to be using manners and shit."

For the second time since I'd been at Carville, my hand had been rejected. I wouldn't make that mistake again.

CHAPTER 8

After work the following day, I returned to my room to find Mr. Flowers, a tall black man in a cowboy hat, standing in the doorway. Flowers was the case manager for our dorm, which meant he had complete authority over us, including our release dates, security level, recommendations for halfway house, and approval for home confinement. The white inmates in our dorm hated him. Most called him "*Le*Roy Rogers." Flowers motioned with his clipboard. He and Doc were engaged in a heated discussion. Mr. Flowers said that all federal inmates who were not U.S. citizens were to be deported, and, according to his records, Victor Dombrowsky was born in Portugal.

"I wasn't born in Portugal," Doc told him.

Flowers insisted the prison records verified his Portuguese birth. If Doc was lying, he said, he would make it a personal priority to see him deported immediately.

"Fine by me," Doc said. "I was not born in Portugal."

Nothing seemed to bother Doc. He put his hands behind his head and lay back on his cot. Flowers gave him a long stare, tapped his clipboard, and walked away.

"What was that all about?" I asked.

Doc shrugged and picked up a medical journal. He was a reading machine. Night and day, he spent every spare moment with his medical publications.

"Where were you born?" I asked.

"Russia," Doc said.

Doc's Russian heritage, in a roundabout way, had led to his imprisonment. As a bright, young, bilingual physician, Victor Dombrowsky was hired by the U.S. government to translate Russian medical documents into English. In an obscure paper, Doc found a detailed account of a chemical used by the Russian army. It was not a compound to be used against enemies; it was used on the army's own soldiers. The chemical, DNP, when ingested, kept Russian soldiers warm in battle during winter months. It increased the soldiers' body temperatures by three to four degrees. They could bear the bitter cold while the enemy froze to death or retreated.

A tiny footnote, buried among the documents, had caught Dombrowsky's attention. It was a detail any other translator might have overlooked. The chemical had an unusual side effect. In addition to raising the body temperature, the drug elevated the basal metabolic rate of the soldiers. The result: dramatic weight loss.

Doc held both medical and pharmacy degrees, and he had dabbled in drug development, for which he had earned a number of patents. He never forgot what he'd read about in the Russian archives. Two decades later, in the mid-1980s, Doc opened dozens of clinics that specialized in weight loss. He treated obesity with a "heat pill." Its primary ingredient was the formula used by the Russians. The advertisements promised that patients would lose up to fifteen pounds per week with no exercise.

"Fat women loved it," Doc said.

Everything went well for Doc until his financial adviser, the man he considered his best friend, became a government informant. The man wore a wire for almost two years, gathering evidence for the federal investigators. He exposed Doc's offshore accounts. He even recorded Doc talking about how best to evade taxes.

"The fucking rat told the bastards where every penny was stashed."

"So you're here for tax evasion?"

"Not exactly," Doc said. "The FDA got involved."

It turned out that DNP, the primary component of Doc's heat pill, had been outlawed as a drug in 1938, though it was still used as a weed killer. Prescribed in the 1920s as a weight loss drug, DNP had caused skin rashes, cataracts, and other medical problems, including loss of the sense of smell. A Viennese physician who wanted to achieve rapid weight loss had taken large doses and literally cooked to death from the inside out. DNP's side effects were the catalyst for the federal Food, Drug and Cosmetic Act of 1938.

Doc insisted there were no permanent side effects to his heat pill because he supplemented the DNP with hormones. In the doses he had prescribed, a patient's body temperature would level off at about 101 degrees. "Some of our larger patients perspired a lot," he said. "A small price to pay."

With DNP outlawed and with Doc's clinic serving Medicaid patients, prosecutors tacked on Medicaid fraud charges.

"How much Medicaid fraud?" I asked.

Doc said that depended on who you listened to. Doc's lawyers argued that $40,000 might have been fraudulent. The U.S. attorney's conservative estimate ranged between $15 million and $37 million.

"Good God!" I said. "How much time did you get?"

"Fifteen years," Doc said, calmly. "The system's fucked. Like *I* could really get a trial of my peers." Doc insisted a jury of physicians would have understood that his heat pill wasn't a drug violation; it was part of a total treatment. And according to Doc, it fell under the FDA's Investigational New Drug exception.

The U.S. attorney threatened to prosecute Doc's children, who were peripherally involved in the business. Doc agreed to a plea deal. He expected a five-year sentence. When Doc appealed the judge's fifteen-year sentence, the U. S. attorney described in dramatic fashion how Dombrowsky had stolen nearly $40 million by prescribing weed killer to desperate, obese women and harbored the profits in the Cayman Islands.

Doc lost the appeal and was stuck with the sentence. But he was determined to spend the time wisely, learning everything he could about medicine, devouring every medical paper published in America, planning the launch of his impotence cure.

My grandfather Harry, who tried to teach me the value of a dollar, and me.

CHAPTER 9

The guard who had caught me talking to Ella gave me an additional daily task—mopping the inmate cafeteria. My instructions were to move all the tables and chairs, nearly three hundred, to one side of the room, mop the empty side, and then move all the chairs and tables to the opposite side to complete the job.

The cafeteria floor was checkered linoleum, so I outlined a ten-by-ten square grid with the mop, careful not to go outside the lines. Then I covered the interior squares with soapy water. To divide the floor into small jobs, sets of perfectly square, manageable quadrants, was satisfying. I have a minor, but not debilitating, obsession with symmetry. Once a square was evenly covered and cleaned, I could admire a job well done, then move on to the next.

The repetitive motion, the back and forth of the mop, was strangely relaxing. The job required no mental energy, and my mind wandered.

My pay for this work was fourteen cents an hour. The meager wage reminded me of my grandfather Harry's attempt to teach me the value of a dollar. When I was four years old, we began a Saturday morning routine. As we would drive to downtown Gulfport, Mississippi, he would remind me of my budget. I could spend exactly one dollar on a toy.

"What if I find something I really want?" I would ask. "And it's a little bit more than a dollar?"

"You'll have to wait," my grandfather said. I would need to save the dollar from this week and add it to the dollar I would get next week, he explained gently, hoping the lesson would sink in.

We would park in front of the old gray post office. Men and women darted in and out of the department stores, restaurants, and shops. My grandfather would introduce me to his friends as we passed on the sidewalk or met in the aisle of a store. When my search for a one-dollar toy ended, we would find a seat at a drugstore counter and order a malted milk shake or french fries. Our last stop every Saturday was always the same. My grandfather would take me to Hancock Bank. It was where everyone in our family banked.

We would wait in line on the black-and-white marble floor with the other customers. When it was our turn at the teller window, my grandfather would prop me up on his knee and introduce me to the teller, as if he knew this bank and its employees would play a vital role in my future. Then he would say, "We need to check on our money." The teller would walk away to check the balance in my college savings account. Upon her return, she would present us with a handwritten balance. I would look at the figure and read it aloud to my grandfather, who, when I got the balance right, would nod and smile.

I was glad my grandfather wasn't alive to see me in prison. I would be ashamed for him to see me like this. But I was determined not to get beaten down by incarceration. I wanted to hold on to, even hone, my skills so I could start a new publishing business when I was released. With a felony conviction, I might have trouble getting hired, but there was no law that prohibited felons from launching magazines. Doc was planning his future, and I would do the same. I had a year to plan, to design mock-ups, and to create a great business and marketing prospectus. I would use this time to plan my financial resurrection. If I could repay my bankers, investors, and creditors— give them a return on their money—it would feel like the money had been invested instead of lost. And I could be back on top.

In the middle of my fifth square, as I neared the five hundred mark of small tiles mopped and cleaned, a guard walked into the cafeteria.

The guard yelled out, "Too slow! You're too slow, inmate!" I nodded, picked up the pace. "You ain't got nothin' comin'," he said.

This is what guards told all the inmates. *You ain't got nothin' comin'.*

I'd heard it every day, but I figured anyone who worked as a prison guard didn't have much coming either, so it really didn't bother me. But I hated to be called "inmate." I thought about asking him to call me by my name. I also wanted to ask why he never woke Jefferson and the prisoners from their slumber in the cooler. But I just mopped.

I thought about my days of investigative journalism in Oxford, Mississippi. At twenty-four, I had launched an alternative newspaper to take on the small-town establishment. Though I garnered lots of accolades, I offended a number of politicians and powerful business-people. But I did have one very important champion: Willie Morris.

The former editor of *Harper's* and author of *North Toward Home*, Willie was the writer-in-residence at The University of Mississippi. He took great interest in my fledgling career as a newspaper editor and publisher. Not so much because he was impressed with my efforts, but because he had a huge crush on my mother, who had recently moved to town fresh from her third divorce.

Willie would call our newspaper offices late at night, inebriated, after the bars had closed and the stores had stopped selling beer. "Mister Editor," he would slur. "For a mere six-pack of chilled beer, I will pen a piece for your fine paper on the ten greatest dogs I have ever known."

I would stop working on the newspaper and take a six-pack to Willie's home on Faculty Row. His guests partook of the beer while Willie sat at his dining room table and wrote out his piece with a black felt-tipped pen on white legal paper. Sometimes, to avoid interruptions, he would put the telephone inside his oven. When he finished, Willie would stumble over, teary eyed at his own prose, and hug me. Then he would insist I join the group for a beer and give him an update on my mother.

After acquiring Willie's dog story and another piece on the greatest Ole Miss football players of all time—paid for, in full, with twelve cold beers—Willie forever greeted me as "Mister Editor." I believed it legitimized my place as a journalist.

As I finished mopping the first side of the room and dragged the tables and chairs from the other side, I remembered Willie inviting his famous writer friends to Oxford. He seemed more than happy to introduce them to me, especially if my mother came along. Alex Haley, William Styron, and George Plimpton all visited Willie. He knew I was a Plimpton fan. Not so much for his literature, but for his participatory journalism.

I'd always dreamed of being an undercover journalist, secretly documenting conspiratorial practices and exploring hidden worlds. Willie arranged for me to meet Plimpton, to interview him for my newspaper, but it was a pretext. For me, the meeting was personal. I wanted to know everything he knew about immersion into a strange culture, clandestine reporting, and impersonation. I wanted to know what it felt like to go undercover, to write about things no one has any business knowing.

I asked Plimpton about his thirty-yard loss in a preseason professional football game when he posed as a quarterback with the Detroit Lions. And about his short stint in professional boxing. And, of course, about his astounding April Fool's hoax in *Sports Illustrated* when he convinced the magazine's readers, and most of the sports world, that a major-league pitcher who had studied ancient Eastern techniques would change the game forever because he had learned to throw a baseball over 120 miles per hour.

Now, as I mopped the cafeteria floor, a hundred checkered blocks at a time, I imagined what Plimpton would do in my place. And it was obvious. He would write about it.

With mop in hand, I decided. I would not be a federal convict. I would simply *pretend* to be an inmate. I would record the stories of the leprosy patients, the convicts, the actions of the guards, and the motives of the Bureau of Prisons. I would uncover why the government decided to experiment with mingling inmates and lepers. And if one of us came down with the disease, I would have the documentation for an exposé.

This was a great plan. This was precisely what I needed to do. As a participatory reporter, I could earn respect. When the guards called me "inmate," it wouldn't matter—it would be my cover. I would play the role, but spend my days listening to and befriending other inmates and, from a safe distance, interviewing any leprosy patient who would talk to me.

Suddenly I felt as if I had escaped. I imagined myself on a stage accepting a Press Club award for the magazine piece I had written about my astonishing adventure. To a standing ovation, I would reach the podium, modestly trying to quell the applause before regaling the audience with my spectacular tales of courage and compassion, bravery and sadness, grit and heroism. That moment would naturally lead to magazine features, newspaper reports, and radio interviews. And, ultimately, perhaps a television special.

I put away my mop and bucket, returned the tables and chairs to their rightful spot, and admired the spotless floor. And I realized I no longer wanted to be transferred. Obviously, I was here for a reason. I was in a remarkable place—one beyond the reach of George Plimpton, even. This was the perfect plan. And I knew the perfect place to start.

CHAPTER 10

The prison library occupied two rooms in a building in the far back corner of the colony. A prisoner in a khaki uniform sat at a desk flipping through index cards. I asked if he could help me find some books on leprosy.

"What size are they?" he asked.

I shrugged, baffled.

"Then you're in trouble," he said, pointing to the shelves. "The library is organized by book size."

Small paperback books occupied one set of shelves. Larger paperbacks covered most of two other walls. Hardbacks lined the windowsill.

"You've got to be kidding," I said.

He went back to organizing his cards. "Warden likes things orderly."

I looked around the room. The other inmates seemed perfectly content with a library that ignored author, subject matter, the English alphabet, and the Dewey decimal system. I moved to the second room, the prison law library, where three manual typewriters occupied a table next to a set of *The Federal Code* and a few shelves of reference books. A small man with wavy red hair was typing diligently. He was focused and fast. He looked up, and I asked if he could help me find some books.

Frank, as he introduced himself, told me that a warden in Texarkana started the movement to organize books by size, and it caught on in other prisons. Frank and a couple of other guys came in when no one was working the desk and, clandestinely, organized the books alphabetically by author, albeit within the parameters of size.

"It's problematic for title and subject matter," he said, "but if you know the author's name, you can usually find the book." When I told him I was interested in books on leprosy, he cringed. "Disgusting," he said, shaking his head again. Then he led me to two books written by residents of Carville, as well as two reference books on the subject. Frank gathered his papers and left. One of the other inmates in the law library asked me if I knew what Frank did on the outside.

Frank, he said, was Jimmy Hoffa's lawyer.

I was intrigued. Hoffa's lawyer would make a perfect interview subject. He would add spice to the exposé, especially if I could convince him to divulge something about Jimmy Hoffa that no one else knew. I sat at a round table and had just begun to read when I recognized Link's laugh from the hallway. Link had a new hobby—following me around, asking questions and making fun of my answers.

"Where Clark Kent at?" he yelled. "I got something for him." Link came in, followed by a couple of his buddies who never seemed to speak. "You made magazines on the outside, right?" he asked. Link laughed and looked at his friends. I told him I had been a magazine editor and publisher. Then he tossed a pornographic magazine onto the table, open to a particularly graphic spread of a man and woman performing a sex act. Link pointed to the publication and asked, "You make *them* kind of magazines!?"

"Of course not," I said. Most of the other inmates in the library were laughing now, too. This was beginning to feel familiar. I took a closer look at the magazine. The pale white woman with jet-black hair had recently had stitches removed from breast enhancement surgery. The scars were pink and swollen and obvious. She also had a vicious rash on her derriere. The man had grease under his fingernails, and the bottoms of his feet were filthy. His mouth was wide open, as if in ecstasy. A handful of teeth were missing.

I pointed out these shortcomings to Link.

"You is lookin' at the wrong parts!" he said. This time, even the guy working at the library information table laughed. "Man," Link said, "you is the borin'est person I ever met in my life!"

I conceded that I probably seemed boring to him.

"What'd you and your old lady do for fun on Saturday night?"

"I don't think it would interest you," I said.

"Come on, man. We ain't got nothin' but time. What you did on Saturday nights?"

"Well . . . sometimes we'd get a babysitter and go out for dinner. Maybe catch a show. Sometimes we'd go to parties."

"They have crack at them parties?"

"No. These were social events. Wine. Beer. Maybe mixed drinks."

"You not only the borin'est person in the world, you is the whitest man I ever met. You was the motherfucker they was talking about when they invented the word *honky*. You white to the core."

I gathered my books and went over to the desk to check them out. The clerk at the table handed me a form to fill out. Link watched over my shoulder, then grabbed the form from the clerk.

"Goddamn!" he said. "Even the motherfucker's name is White!"

Carville's two miles of covered corridors were built because it was believed the sun aggravated symptoms of leprosy.

CHAPTER 11

After work each day, I walked the perimeter of the prison side. Walking, even in circles, made the time pass. As I made my way around the corridors, I passed by patients and inmates. Best I could tell, about one-third of the inmates were white, one-third black, and one-third Hispanic. The number of healthy inmates (we were called *work cadre*) and medical inmates (called *broke dicks* by the healthy inmates) was about even. Work cadre inmates wore green uniforms. Medical inmates wore khaki. Over one hundred medical inmates were in wheelchairs. Combined with the leprosy patients, more than two hundred wheelchairs moved around the colony on a regular basis. The patients, except for Ella, owned motorized wheelchairs. Inmates had the basic kind, pushed by humans.

My private walks were often cut short by Link. He had yet to get a job assignment, so he spent his days wandering the colony. "This place is a motherfuckin' country club!" he yelled. "I ain't never leaving this place!" If Link saw me, no matter how far apart we were, he would scream as loudly as he could to get my attention. "Clark Kent! You borin' motherfucker!" Some of the older prisoners demanded that the guards assign him a job because he was disturbing their peaceful mornings.

Link had spent time in other prisons, so I imagine Carville did feel like a country club. Six television rooms carried basic cable, as well as HBO. The recreation department had a pool table, a Ping-Pong table, an arts and crafts room, a music room, and a television dedicated to Nintendo. Outside there was a horseshoe pit, a shuffleboard court, a sand volleyball court, a walking track, four handball courts, a full-sized basketball court, stationary bicycles, and a fully equipped weight room.

Link started making himself at home in my room, much to Doc's chagrin. Link's nonstop talking distracted him from his reading. If I happened to miss Link's visit, Doc would say, "Your friend honored us with his presence again."

Confounded about why I spent time with him, Doc asked, "You find these people entertaining?"

Link was entertaining, but that wasn't the only reason I spent time with him. Link told everyone—inmates, guards, even the leprosy patients—that I was out of place. He reminded them I was an idiot criminal who forgot to keep any money. He imitated the way I greeted people, how I apologized when I bumped into someone, how nervous I acted when someone around me violated a rule. In his own way, he was saying exactly what I wanted them all to know. Link told them I was different. He made certain everyone knew I didn't belong here. I didn't have to say a word.

And even though Link was making fun of me when he called me Clark Kent, I enjoyed being likened to a superhero.

When I was five, *Underdog* was my favorite cartoon. In my parents' bathroom in front of the large mirror, I would tie a towel around my neck and flex my muscles like the superhero.

I loved to watch Shoeshine Boy run into a phone booth, transform into Underdog, and save Sweet Polly Purebred from the evil villain. On the rare occasion when Underdog was close to defeat, in need of extraordinary powers, he would open the secret compartment of his ring and recite a rhyme: "The secret compartment of my ring I fill, with an Underdog Super Energy Pill." When he swallowed the red pill, Underdog became powerful enough to move planets.

On a Saturday morning, I searched through my mother's medicine drawer and found what I was looking for—a secret energy pill. I pushed the tiny pill out of the flat, plastic container, put it in my pocket, and walked down the block to my friend Mary Eliza's house. Her mother greeted me at the back door. "That's a fine cape, Neil," she said, "I'll get Mary Eliza."

Her mother poured us each a glass of apple juice. I turned up my glass and gulped it all down without taking a breath. I let out a sigh and put the glass down on the table, like I had just won a drinking contest.

In the backyard, I put my hand next to Mary Eliza's ear and whispered that I was going to fly. Mary Eliza had the highest tree house in the neighborhood. I tightened the knot in my cape and took off up the ladder. I climbed up to the tree house and then pulled myself up onto the roof.

I stood at the edge of the shingles and looked down. The tree house was high, far above the basketball goal. Too far to fall. I needed to fly. In my pocket, I found the pill and bit into it. It was bitter and dry. I swallowed all that I could and moved my toes over the edge of the roof. I bent my knees and let my arms relax. I didn't count. I had a cape, a secret energy pill, and the full belief I could do anything I put my mind to.

I swung my arms out and dove off the tree house roof.

In the doctor's office, my mother told the nurse about the accident. Then she whispered, "And he took one of my birth control pills."

The nurse gave me instructions for a urine sample and handed me a plastic container. In the men's room, I filled the cup. I wished it had been bigger. I could have filled it three times.

"My!" the nurse said, carefully taking possession of the overflowing container.

"I got more if you need it," I bragged.

Back in the waiting room I told my mother I could have filled a much bigger container. She kissed the top of my head and said, "You were put here to do great things. Don't ever forget that."

Jimmy Harris

CHAPTER 12

My plan to write an exposé about the convicts and the leprosy patients fit perfectly with my mother's early vision for me. I had an opportunity to turn this incarceration into a great piece of journalism. And the next thing I needed to do was get more of Ella's story. I had seen her in the hallways and a few times in the patient cafeteria, but each time a guard was close by. I would have to catch her alone when we both had plenty of time to talk.

In the meantime, I had met another leprosy patient, Jimmy Harris, who talked nonstop. He had been sent to Carville in 1938. He was writing his own book about life in quarantine and was happy to talk about it to anyone who would listen. He had already picked a title, *King of the Microbes*. Jimmy had a head of thick, white hair and a slight curvature of the spine. Except for a claw hand, he looked perfectly normal for someone eighty-two years old.

Jimmy discovered his own leprosy on a summer evening in 1937. His father was hosting a barbecue. Later that night, as Jimmy undressed and removed the short pants he had worn that day, he noticed a spot on his leg. A clean spot. Although the rest of his leg was covered in fine Texas dirt, none adhered to the oval on his thigh. Jimmy knew the signs of leprosy because his brother, Elmer, had been stricken with the disease four years earlier.

"You lose your ability to perspire," Jimmy told me with great authority. "I knew what I had, but I went to see a doctor anyway. Took my 1931 Chevy Coupe over to Beaumont. That sucker would fly." Then he went on to talk about horsepower and cylinders and other mechanical stuff. He seemed more interested in automobiles than leprosy.

I made notes on Jimmy's stories, and I recorded conversations I overheard in the cafeteria between the other leprosy patients. They called themselves "secret people." I felt like a voyeur listening to them talk about their confinement, misdiagnoses, lost families, and the heartbreak of love affairs. What I didn't hear in the cafeteria, I discovered in the books from the library.

In the 1850s, Carville was called Island, Louisiana. The land was owned by Robert Camp. Indian Camp plantation, as it was known then, was a successful sugar plantation, but it was abandoned when Camp lost his fortune after the Civil War. The plantation sat in disrepair, unoccupied for thirty years, before the State of Louisiana leased the land in 1894. The 360-acre plot, along with a decaying manor house and slave quarters, was then designated as the Louisiana Leper Home. After that, all lepers in Louisiana were sent to the remote colony. The geography was perfect for outcasts. The plantation was virtually impossible to reach by land: a washed-out road with no outlet, leading to a tiny drop of land that looked like gravity had pulled it into the river's path. It was known primarily to boat captains who navigated the sharp 180-degree turn in the Mississippi River just south of Baton Rouge.

In the early days, doctors and nurses were reluctant to come to the home. There was no running water, little sanitation, and no budget for improvements. The first residents shared the buildings with snakes and bats.

A compassionate physician from Tulane who had studied leprosy traveled to the East Coast to recruit an order of nuns to provide care for the leprosy patients. In 1896, the first of the Sisters of Charity arrived at the colony.

A series of bizarre events that began in 1914 altered the course of the colony's history and led to its establishment as the national leprosarium. John Early, a veteran of the Spanish-American War, had been diagnosed with leprosy in 1908. A native of North Carolina, Early had been quarantined in a series of temporary posts. North

Carolina sent him to Washington, D.C., where he was imprisoned in a tent on the Potomac River. Officials in Washington transferred him to a colony in Massachusetts, but authorities there refused to accept him. He spent years moving among boxcars, ramshackle houses, and jails. No state or territory would permanently accept him within its province. In late 1914, after six years of provisional quarantines, Early checked himself into the fashionable Hotel Willard in Washington, D.C., where it was rumored the vice president and several senators had also established residence. He invited the *Washington Times* newspaper editor and other reporters to gather in his room at the Willard. Early guessed that if he mingled among the powerful, if he demonstrated in a very public manner just how easily one of the four hundred lepers at large could walk among healthy citizens, a national home for victims of the disease would not be far behind. It took several years of public appearances for Early's dream to bear fruit. But in 1917, Woodrow Wilson signed legislation authorizing $250,000 for the care and treatment of people afflicted with leprosy. They would be transported free to a yet-to-be-established U.S. leprosarium. Four years later, Carville was chosen.

Before the establishment of a national leprosarium, lepers were at the mercy of local officials or, sometimes, a renegade posse of men trying to protect their families. Some who contracted the disease were segregated from the general public in jails or dilapidated homes known as pesthouses. Not unlike the bells and clappers of biblical times, yellow flags or large quarantine signs were attached to the pesthouses to warn citizens of the danger of infection. Those who weren't forced into quarantine lived in fear of mobs that threatened their families with arson. When a family member was diagnosed with leprosy, unafflicted relatives often fled to avoid neighbors who would ostracize spouses and children. There was widespread misunderstanding. Many still believed that leprosy was a disease of the soul, that the victim had been stricken by God for misdeeds.

Early intended to secure a home for himself and others with leprosy, but it brought about an unintended consequence: a new national policy of segregation tantamount to imprisonment.

Even though the new policy of mandatory confinement came from federal law, it was enforced by local officials. Enforcement played out differently depending on the fears, biases, and misunderstandings of state and county law enforcement officials. Some individuals who contracted leprosy were brought to Carville in shackles. Others were locked in jail cells until paid couriers, sometimes armed with handguns, could be hired to transport them to the leprosarium. In some states, if multiple family members were afflicted with leprosy, their home might be torched. An infected child's toys, clothes, and books were incinerated. Parents were forcibly removed from their children. Children were pulled from the arms of their parents.

Fear of the disease was so rampant that arrivals at Carville took on new names to protect their families from the stigma. A gentleman named Stanley Stein, who contracted the disease at age nineteen and was later confined at Carville, wrote that he felt like "an exile in his own country."

If there were any bright spots in the early history, it was the Sisters of Charity. The nuns were dedicated to the physical and spiritual comfort of the outcasts. According to patients' accounts, the nuns' kindness was the saving grace of the colony. One sister who arrived at the turn of the century told the patients she would never use the term "leper" to describe them. Instead she called them "my friends."

Over the decades the residents at the leprosarium sponsored Mardi Gras parades, launched their own publications, organized a patient federation, Boy Scout troops, a softball team, and even a Lions club. The stories amazed me—as did the history of the leprosarium. The more I learned, the more intrigued I became with this extraordinary community of people who established their own rituals and traditions.

But the disease itself was even more evocative. One of the reference books that piqued my interest was an illustrated medical atlas with more than a hundred clinical photographs. I carried it with me around the colony, but I was careful about when and where I studied it. I kept it hidden between two notebooks.

In graphic and shocking candor, the photographs depicted the effects of leprosy at progressively severe stages. The patients at the beginning of the book, the ones identified as having the most immunity, looked relatively unscathed—patches of skin with pigmentation loss, eyebrows with no hair growth, minor abrasions. As I turned the pages and moved into the section that depicted the borderline diagnoses, I saw pustules and swollen faces, lumpy protrusions on the ears and forehead, torsos covered with red lesions. At the back of the manual, the images became more and more horrific. The patients with little or no immunity, a diagnosis called *lepromatous leprosy,* had feet so deformed and twisted I couldn't tell if the toes had disappeared or if they had been grafted together. Noses were virtually nonexistent. Appendages were so disfigured that amputation might have been preferable. And patients suffering from a rare form of leprosy called *Lucio's phenomenon* were left with huge holes in their flesh that looked like their skin had been eaten away to the bone by parasites. Their faces were scarred, consumed to the point of looking inhuman.

Black boxes were superimposed over their eyes to protect their identities, but Carville was home to the very last leprosy patients in North America. I assumed some of the patients here must have been featured in this book.

But the most worrisome revelation about leprosy—confirmed by Doc and the reference books—was that no one was certain how the disease was transmitted, no vaccine existed to prevent the spread, and no test was available to determine who was naturally immune and who was susceptible.

CHAPTER 13

My menu board illustrations had become popular with the leprosy patients. They especially liked my President Clinton caricatures. So I added more colorful illustrations and entertaining slogans to each day's board. When the meal included Cuban Chicken, I sketched a portrait of Fidel Castro smoking a cigar, holding a chicken by the neck. On Mexican Day, I designed the text to fit inside a large sombrero, and I added Taco Bell's slogan, "Make a run for the border." The inmates and leprosy patients thought it was pretty funny; it didn't amuse the warden nearly as much.

To get a jump on my early morning duties, I started to transcribe the menu board in the patient dining room every day after lunch. As I wrote on the board one afternoon, I heard a voice behind me announce, "Hey, they finally gave us one who can spell!" A gaunt black man wearing a hat and a khaki coat reached out to shake my hand. "I'm Harry," he said with a crooked smile, "nice to meet you." He looked like the man I saw waving through the screen on my first day. His hand had only two appendages—a complete thumb and part of an index finger. His other digits looked like they had been absorbed or maybe burned off. He seemed friendly. I didn't want to reject him, especially since my own hand had been twice refused. But I didn't want to touch him either. And how would I do it? Grab his finger and shake? Put my open hand between his thumb and index finger and let him grab on?

He noticed my hesitation. Harry's smile disappeared. He put his hand back in his coat pocket. "You've got real neat handwriting." He smiled again and told me to have a nice day. I looked around the room and saw the other patients. They exchanged glances and shook their heads. As I watched Harry walk away, I knew I had hurt his feelings. I wanted to call him back, apologize, and accept his hand.

But it was too late.

Little Neil, Maggie, and me on the visiting room deck.

CHAPTER 14

"May I *please* borrow your iron?" I asked again.

CeeCee had no intention of giving up her iron. CeeCee was a federal inmate, too, but she insisted that we use feminine pronouns when speaking to, or about, her. CeeCee's shirt collar was turned up around her thin neck. The top four buttons on the tight green shirt were left open to reveal what would have been cleavage had she been a woman. I guess she figured our deprived imaginations would fill in the rest. CeeCee had altered her prison-issued slacks to look like capri pants.

"Nobody touches my iron," CeeCee said, with a hand cocked on her hip.

My shirts were wrinkled, and Linda and the kids would arrive in the visiting room in less than an hour for our first visit.

CeeCee offered to iron my shirts. She said she would charge twenty-five cents per shirt, and she promised to have one pressed and ready by the time my family arrived. She bragged that she could iron pants better than a professional dry cleaner, that her shirts held up for days, even in the Louisiana humidity, and that she would even sew a Polo man onto the left breast pocket. "I can add starch," she added.

I loved starch. I had always requested heavy starch at the cleaners. My shirts and pants were as stiff as cardboard. Starched clothes looked old-fashioned and stable. My clothes reminded older businessmen around town of a simpler, more genteel way of life. And I liked that.

I paid CeeCee in quarters and asked her to hold off embroidering the Polo man.

Two weeks had passed since Linda dropped me off at the colony gate. Anxious, in my perfectly pressed prison uniform, I stood in the hallway with the other inmates. We gathered behind a barricade the guards set up on visiting days. As families arrived, a guard would escort us to the visiting room to reunite with our wives and children.

Linda was late. After waiting about an hour, I started to worry. Maybe she'd had a hard time getting the kids ready by herself. I couldn't help wondering if something had gone wrong. An accident. A flat tire. A family emergency. I felt helpless. I did the only thing I could do. I waited.

One other inmate had been waiting just as long. He said I shouldn't be disappointed if my family didn't show up. Sometimes, he said, they just can't face another trip to a place like this.

The guard finally called my name. As I walked to greet Linda and Neil and Maggie, I hoped my children wouldn't be shaken by seeing me in jail, as an inmate, in a prison uniform. I didn't know how I would explain an inmate cursing or stealing. Or, God forbid, if they encountered a leper. How would I explain these things to a six-year-old and a three-year-old?

In the visiting room, my inmate friends were already settled in with their families. Doc sat at a round table in deep conversation with his girlfriend, a former nurse. Chief, a Native American inmate with long, silver hair, played a card game with his grown son. Steve Read, an airline entrepreneur, held hands with his wife, who looked like she was ready for a fashion shoot. Other families played dominoes or cards, like they had nothing better to do with their time.

My family stood next to the toddler play area. Neil and Maggie ran toward me. I bent down and hugged them both tight. They yelled that they wanted to go down the slide on the playground. I hugged Linda, and she gestured for me to go outside with the kids. She would wait on the deck. "We can talk later," she said.

Maggie, Neil, and I played a makeshift game of baseball using imaginary bats and a racquetball, but the kids quickly grew bored. Little Neil had a better idea. He had been through one judo lesson in New Orleans. He grabbed my shirt and tried to flip me. I fell into his move so that I ended up on my back, sprawled on the grass.

An armadillo waddled toward the playground like a tiny prehistoric creature protected by nine bands of armor. It sniffed around the fence and then wandered off toward a ditch. I noticed the leprosy patients were out enjoying the morning too. Harry rode his bike along a concrete path. An overweight patient passed us in his cart driving toward the golf course. A woman with one leg zipped by in an electric wheelchair. For the most part, the patients ignored us. But one stopped and stared. Anne. She looked in our direction and stood with her hands on her hips. I struggled up from the ground and wiped the grass from my pants.

In her midsixties, Anne had a narrow face and dark circles under her eyes. Her hair was dyed black. When she was first quarantined at Carville, away from her family and alone, she fell in love with one of the other patients. A few months later she discovered she was pregnant. Anne knew the policy. Everyone did. It was simple, and cruel. Babies born to women with leprosy were taken away immediately and put up for adoption. It was possible for a quarantined woman to make arrangements with her extended family to care for the child, but like so many at Carville, Anne had been disowned. As far as her family was concerned, she was dead. No one in her family would take the child.

On the day Anne's daughter was born, one of the nuns, a Sister of Charity, assisted with the birth. She washed and weighed the child. Then she wrapped her in a blanket and placed her in the wicker basket the Sisters had used for decades to transport the children born to leprosy patients.

Anne called to the Sister, "I want to see my baby."

In the room next door, the nun collected the baby girl from the basket. She stopped at the doorway of the birth room and held the baby up high.

"I want to hold her," Anne said, but the rules were clear.

The nun cried as she pulled Anne's baby next to her own breast and turned away.

Just outside the prison visiting area, Anne watched as dozens of imprisoned fathers played with their sons and daughters. She was confronted with all she had missed. Even convicts were allowed to hold their children.

One of the other inmates on the playground noticed me watching Anne. "Surreal, isn't it?" he said.

I nodded and looked over at my children. Neil practiced his judo moves on Maggie, who seemed perfectly happy to be flipped onto the ground and repeatedly fallen upon.

"First visit?" the inmate asked.

I nodded.

"Always tough," he said, "especially on the wives." His own daughter was seven. He considered himself fortunate. Inmates with teenage children, he said, had a more difficult time. Older kids seized on the fact that their fathers had broken the law and, consequently, had no right to tell them what to do. "Young ones don't know any better," he said. "They figure all daddies do a little time."

I warned Neil to be careful flipping Maggie. She jumped up and said, "Daddy! I know judo, too!" Then she lurched forward and delivered a head butt directly to my crotch. Maggie smiled. She was proud that her move had bent me over. I patted her on the head and told her I thought she had a great future in the martial arts. One of the other kids asked Neil and Maggie to join them on the pirate's ship.

I saw Linda on the wooden deck. She sat alone and stared out at the colony. She looked sad, and it was my fault. Neither one of us wanted to be here. And we were both at a loss for words.

Maggie, me, Linda, and Little Neil (left to right) *in the courtyard at St. Peter's by-the-Sea, Easter 1992.*

CHAPTER 15

I met Linda in Oxford, Mississippi, in 1984 when we were both hired at Syd & Harry's, a new restaurant that promised to change the way the small town experienced fine dining. We were both twenty-three years old. Before I met Linda, I had dated pageant girls. In Mississippi, they are everywhere: young women who spend an inordinate amount of time on makeup, hair, exercise, fashion, diet, and posture. I endured constant talk of calories and minute weight gain because the girls made me look good. To walk arm in arm with a girl who was always onstage, aware—or at least imagining—that people were constantly looking, did wonders for my image.

Linda was different. She was as attractive as any of the pageant girls, but she would never have degraded herself by allowing beauty pageant officials to judge her appearance. Her light blond hair perfectly framed her taut olive face. Her nose formed a flawless angle above thin lips that revealed a bright and spontaneous smile.

Her loveliness matched her demeanor. In everything, moderation. A bit of each day spent outside with family, with a book, at a good meal, with a glass of wine—at the most, two. At the best restaurants, she would take four or five bites of the entrée and lean back, satisfied. She would say it was one of the best meals she could remember. Of course, I would finish mine, and eat the rest of hers, too.

When we met, Linda was a graduate student in literature and creative writing. I had assumed she'd come from a home where education and enlightenment had been handed down through generations.

But she came from a family of Mississippi farmers. She was the first in her family to graduate from college.

Her relatives had treated me like blood kin. Her grandfather, a dirt farmer who amassed a fortune over the years, invested $10,000 in my newspaper business. After my terrible stewardship of his money, and everyone else's, he spent another $5,000 to help retain my criminal lawyer. He kept no ledger when it came to family. There were no debts. Only gifts. When I asked how I could ever repay him, he said in a soft southern drawl, "Do the same thing for your children." As I awaited my criminal sentence, he pulled me to the side and asked, "Does the judge have a price?" Then he winked and put his arm around me. We both knew he would never break the law, but he wanted me to know that if he were that kind of man, he would have done it for me.

In my mind, Linda was just about perfect. Attractive. Smart. Petite. Quick-witted. The kind of woman people wanted to be near.

We left Oxford for our honeymoon under an awning of brilliant, yellow ginkgo leaves. That evening at the Ritz-Carlton in Atlanta, with its polished wooden walls and ornate antiques, the clerk handed me two keys. One for the room, the other for an honor bar.

As Linda bathed in the opulent marble tub, I unlocked the bar and tore into the Godiva chocolates and imported nuts. I opened a bottle of Moët & Chandon, and then a bottle of red wine, then a bottle of white, just in case she preferred it to champagne. I scattered M & Ms and caramel squares and imported candies on the bedside and coffee tables. Linda stepped into the room in her robe, fresh from her bath, ready to consummate our marriage. Her skin was smooth and lucent. I could smell the lavender bath oil from where I sat in an overstuffed chair, my mouth full of macadamia nuts. She saw the pile of empty wrappers, corks, and foil. She glanced over at the open bottles and the mounds of candy.

"Look, honey," I said, "they have this great bar in our honor."

Under her breath Linda said, "I've made a mistake here."

We loved to tell the honor bar/honeymoon story. We had laughed about it a hundred times with friends and family. And when Linda laughed without reserve, small dimples on her cheeks exposed a tenderness underneath her regal air. I hadn't seen the dimples in more than a year. Now, on the deck of a prison visiting yard, we didn't have much to laugh about. I sat down next to her and held her hand. "How are you?"

She forced a smile and glanced down at the ground below. At the entrance to the prison, a guard had waved a handheld metal detector over Linda's entire body. Then he did the same with the children. One of the female guards said that Linda's sundress, a dress she had worn to church, was too revealing, too low cut. The guard added that she would be turned away in the future if she arrived in anything like it.

Linda was mortified. She had done nothing wrong. Except marry me. The last year had been a series of humiliations. Bankruptcy. Criminal hearings. Newspaper headlines about my prison sentence. A front-page photograph of the two of us, with Linda prominently featured in the foreground. Expulsion from private clubs where our friends still gathered.

I had convinced Linda early in our relationship that together we were going to make a difference. With her talent and refinement, and my drive and salesmanship, we could conquer any task. I was confident and self-assured—very different from her father—and I knew she was drawn to that trait in me. I asked her to join me in building a publishing empire that would bring with it influence and power, money and prestige, and lead us to places neither could quite imagine. The details of our conquests were not fully developed, but I assured her that the ride would lead in one direction. Up.

We sat among the other inmates. The words came hard.

I told her about Doc and Link and Frank Ragano and CeeCee and other interesting inmates I had encountered. I left out the stories about the leprosy patients. I told her every funny incident I could remember. I hoped to see a glimpse of her dimples. But she couldn't muster much of a smile.

"I'm glad you're having a good time." She shook her head. "I'm sorry. That came out wrong."

But it didn't. She was bearing the brunt of things now. She lived outside. She had to face the stares from neighbors, from other moms in the carpool line. She answered phone calls from creditors. She overheard the talk about how much money I'd lost and about how many people I'd hurt.

"I'm so sorry," I said. "So sorry I put you and the kids in this situation."

Linda was tired of my apologies. I had said *I'm sorry* so many times it didn't carry much weight now.

"If you're really sorry," she said gently, "you'll change."

I knew she was counting on that. If I could change, she had told me, she might be able to take me back. But I didn't know how to change. And, secretly, I didn't think I needed to. I didn't want to become a different person. I simply needed to operate within the bounds of the law. I wasn't about to let go of the skills I'd developed or the plans I'd made. And I had a new plan as an undercover reporter that was going to secure my future. But I didn't tell Linda. At least not yet. No need to get her involved. Plus, she might not have liked the idea.

Little Neil grabbed my hand and pulled me toward the playground. The racquetball had bounced over the fence and stopped about ten feet outside the perimeter.

"Go get it, Dad," Neil said. Maggie stood next to him and pointed at the blue ball. They waited for me to step over the three-foot fence and get the ball. I looked around. A guard was watching me from the deck. It had been made clear to the inmates. *If you go outside the fence, you lose visiting privileges.*

For as long as my children could remember, I had ignored fences and boundaries and rules. I climbed buildings to get balls out of gutters. I jumped curbs to get closer to the entrance of football games. I talked clerks into giving us rooms at overbooked hotels. Nothing much had prevented me from getting what I wanted, and I made sure my children knew it. Now I stood at the edge of a knee-high fence, embarrassed to be so helpless.

CHAPTER 16

That night, just before lights out, Doc asked if I would take a look at the mole he had removed with the razor blade. He stood up, turned his back to me, and pulled up his khaki shirt.

"Did I get it all?" he asked.

I stared at the scab left from Doc's self-surgery. The spot looked black and jagged. I told him I couldn't really tell much.

"Is any of the mole left?" he said. "Can you describe it?"

I couldn't believe he wanted me to examine his back. I couldn't begin to distinguish between the remnants of the dark mole I'd never seen up close and the scab forming over his incisions.

"I can't tell," I said.

Doc sighed in frustration. He walked over to the mirror and looked over his shoulder to try to get a good look at his handiwork.

After lights out, I lay in the dark thinking about Neil and Maggie and how my imprisonment would affect them. Linda told me that she and the kids had been invited to spend the month of July in a Florida condominium with our friends the Singletarys. I thought she needed a vacation from my troubles more than I needed visits. I encouraged her to go.

Linda had never promised to stay with me. The last year had been difficult for her. Too many lies. Too much trust to regain. From the outset, she had never made any guarantees.

I didn't know if Doc was still awake, but I asked aloud, "I wonder if my wife is going to stay with me."

Groggy, Doc answered, "Was your marriage solid?"

We had been seeing a psychologist, I told him.

"*Not* a good sign."

"I hope we can pick it back up when I get out," I said. I thought the counseling had been helpful.

"Don't waste your money."

"What do you mean?" I asked.

"If you have a male counselor, he'll just want to sleep with your wife," Doc said. He rolled over in his bed and pulled the covers up over his shoulder. "If she's female, she'll just want to screw you."

A stained-glass window behind the pews in the Catholic church where leprosy patients worshipped.

CHAPTER 17

On a Sunday morning in late June, in spite of the three-hour round-trip, Linda brought the children to Carville for the Catholic church service. Families entered the Catholic church from the outside. Inmates and leprosy patients entered from within the colony. A long, curved hallway connected the church to the main buildings. About forty inmates waited for Father Reynolds at the doorway. The leprosy patients went straight inside through an automatic door donated by an ancient order of knights whose members had contracted leprosy during the Crusades.

Stan and Sarah, a couple from the Caribbean, approached the church entrance. Both blind, they wore oversized sunglasses to hide their sightless eyes. Stan tapped his white walking stick hard against the floor and walls. Sarah, his elegant wife, who had contracted leprosy in her early twenties, held Stan's left arm in hers. She trusted him completely to guide her through the labyrinth of hallways. The blind leading the blind.

"What a curse," one of the inmates said as they passed us by.

Father Reynolds arrived and led the inmates into the chapel. The church was built in the shape of a cross. The large center wing was available to inmates and families. The left wing was reserved for leprosy patients, the right wing for visitors and nuns. The wings met at a marble altar draped in an ornate cloth. Father Reynolds knelt at the altar. A gold chalice stood at the center of the table.

Linda, Neil, and Maggie waited in a pew in the center wing with the rest of the wives and children. I rushed over and sat between

Linda and Neil; I put Maggie on my lap. I was overjoyed to be in church with my family. As Father Reynolds started the service I put one arm around Linda and with the other pulled Neil closer. No guards loomed over us. Even in prison, it felt like a safe place. A sacred place. I had never been in a Catholic church. I was Episcopalian, but the services were virtually indistinguishable.

During the first hymn, I saw an inmate take a roll of quarters from his girlfriend. Across the aisle, another took a small bottle of bourbon from his wife. As we sang stanza after stanza about the glory of God, an eruption of smuggling—coins, cologne, liquor, and Ziploc bags full of cookies and fudge—ensued between husbands and wives.

As Father Reynolds called for the confession of sins, we pulled out the small pad under the pew and knelt. Father Reynolds read from the missalette: *Let us call to mind our sins.*

The congregation of leprosy patients, nuns, inmates, and families read aloud in unison: *I confess to almighty God that I have sinned.*

I should have been thinking about my own faults and words and deeds, but I couldn't resist watching the other men and women pray. The last time I had been in church, I wore a Brooks Brothers pin-striped suit and a $300 pair of shoes. I had been surrounded by my Episcopal friends—a pediatrician, a real estate appraiser, a stockbroker, a newspaper editor, and dozens of other upstanding citizens. Now, on my knees reciting a prayer for forgiveness, I was surrounded by a different kind of congregation. Nat Sykes of Chardon Insurance had collected millions in insurance premiums and left thousands of Louisiana drivers without liability coverage. Daniel Stephens, a Texas savings and loan president, had hired a professional hit man to assassinate his Thoroughbred horses in order to collect death benefits. Lawrence Daily, a jovial Cajun, had been convicted of a price-fixing scheme to corner the wholesale crawfish market. Steve Read, an airline entrepreneur who had laundered money to keep his regional airline afloat, owned the charter airline that had crashed and killed most of Reba McEntire's band. Ira Kessler collected fees through the front door of his New York pet-funeral business and took money through the back door from Vietnamese restaurant owners. The Kingfish was the first

CEO sentenced to prison for dumping industrial wastes in the Mississippi River. The white-collar criminals were joined by an assortment of drug dealers, as well as the Half-man, an expert lithographer and counterfeiter, who was amputated so far up his torso that he was lodged inside a bucket so he wouldn't fall over in his wheelchair. And there was Sidewinder, who kept his dead mother in a spare room so he could continue to collect her Social Security payments, an act considered so heinous, even among the inmates, he was friendless.

I had been so excited to see my family, I had rushed right past the leprosy patients. I could see them clearly now; they were not kneeling. Some did not have knees. I saw Harry and Jimmy Harris, but Ella wasn't in church. Stan and Sarah, the blind couple, sat in the front row.

The church had not always been a place of comfort for those afflicted with leprosy. At one point in history, anyone deemed leprous was cast out. A "leper's mass" was performed. As the afflicted man or woman watched from the outskirts of a settlement, a priest would preside over the burial, a funeral symbolic of the death of the leper. The priest would pour dirt into an empty grave. Cast out, they wandered alone until death. Occasionally, a leper would find a Lazar house where fellow sufferers gathered and lived.

The men and women who sat across from me now came to the same Catholic church that had once banished them. They prayed in unison: *I ask blessed Mary, ever virgin, all the angels and saints, and you, my brothers and sisters, to pray for me to the Lord, our God.*

At Communion, Father Reynolds invited a group of us to stand at the altar. Linda, Neil, Maggie, and I joined Sister Margie, Steve Read and his wife, and a few other inmates with their families. We held hands and formed a semicircle behind the altar table. Neil and Maggie stood between Linda and me. I held hands with Sister Margie; Linda with an inmate who had been convicted of money laundering.

The leprosy patients remained in their wheelchairs and in their pews. Father Reynolds recited the liturgy, drank from the Communion cup, and broke the bread. In the Episcopal church, the next step would be the shared cup.

I had been a lay Eucharist minister, a fancy title for someone who gets to wear a robe and help deliver the wine to parishioners during Communion. I would put the ornate cup to the lips of the communicant, reciting: "the Blood of Christ, the Cup of Salvation." During the course of a single Communion, more than a hundred people would drink from the cup. Some lathered their lips with saliva before they drank. Others opened their mouths wide like a fish. Women who wore too much lipstick left a thin, oily film on top of the wine. Some of the older people backwashed bits of the wafer into the dark red wine. It took faith to be a lay Eucharistic minister and still drink from the cup.

A part of me felt honored to be breaking bread with the patients, privileged to be in a church that offered solace and comfort in a place that had seen so much suffering. I didn't want to offend any of the patients, like I had Harry, but in good conscience, I couldn't let my family be exposed to the danger of a shared cup. I held my breath, ready to tell Linda, Maggie, and Neil to pass on the cup if necessary.

Father Reynolds stepped in front of each of us with a small plate of wafers. We stood with hands turned up, and he placed a wafer in each palm. I watched carefully as Father Reynolds gave the bread to Harry and Jimmy and Stan and Sarah.

A woman held her mouth wide open, and Father Reynolds placed the wafer on her tongue. Stan and Sarah held out their tiny fingers— so small because the digits had been absorbed into their bodies. Father Reynolds placed a wafer in their palms and then recited loudly, "The Body of Christ, the Bread of Heaven." That was their signal that the wafer was in place. They couldn't see it. And they couldn't feel it.

I watched Harry balance the wafer on what was left of his hands. The light behind him poured in from a dark blue, stained-glass window. Inscribed on the glass were words from the Gospel, *I will console them in all their afflictions.*

Father Reynolds returned to the altar. It was time to drink the wine. He held the chalice high in the air, recited a few phrases, and took a drink from the cup. Then he put it back on the altar and recited the closing prayer.

None of us would be allowed to touch the chalice. I understood why someone with a disease wouldn't be allowed to drink, but I didn't know why inmates were disallowed. Part of me was relieved, but I also understood that I occupied a new place in the eyes of the church. I was an outcast right alongside the victims of leprosy.

CHAPTER 18

On Monday morning, I found myself alone with Ella in the cafeteria. We finished our first cup of coffee before 5:00 A.M. She asked if I had gone to church. I told her that not only did I go, but that my family had joined me. That seemed to make her happy. With no guards around, I asked Ella if she would tell me more about the bounty hunter who brought her to Carville over sixty years ago. With the first hint of sunrise filling the windows of the cafeteria, Ella breathed in deeply as if to pull the memory from a faraway place. She put her shiny hands in her lap and intertwined her fingers. I wished for a tape recorder, but inmates weren't allowed to have such devices; one of the men had figured out how to use the parts to build a makeshift tattoo gun.

"Do you mind if I take notes?" I asked.

Ella shrugged like she didn't understand why I would be so interested. Then, with the air of a seasoned raconteur, Ella continued her story.

The man with the gun drove the truck to Ella's farm and stopped in front of her tenant house, a shack with two rooms and a stove on the front porch. The man walked to the back of the truck, grabbed a hammer and tacks, and picked up the quarantine sign. "Stay put," he told her.

Ella's father was as thick as an oak and tall enough to duck to get through his own front door. He farmed twenty acres of land just outside Abita Springs. Ella's brothers helped with the crops, and Ella had planned to join them the following season. Her father must have heard the bounty hunter's hammer as he tacked the sign to the side of the house. He came in from the field. The bounty hunter looked over his shoulder, watched for a moment, and then turned back to finish his job.

"My daddy come over to the truck," Ella said, "and pick me up." He carried Ella to the front porch and told her to go inside. When the bounty hunter finished, he stared at Ella's father, tall and straight on the porch.

Ella's father spoke first. "She my girl," he said. "I'm takin' her."

The man pushed his coat back, just like he had at the school, to expose his pistol. Ella's father looked down at the gun and repeated, "I'm takin' her." He must have known that this man they called the bounty hunter wasn't a bounty hunter at all. He was hired by the State of Louisiana as a driver. He wasn't even a real law enforcement officer, though he tried to pass himself off as one. He was just a hired hand, paid $10 for each delivery to the leper home.

The bounty hunter held his hand up, fingers outstretched toward the porch, and curled his index finger in against his thumb. "Three days," he said. "If she ain't there in three days, I'll come for her." Ella's father didn't move. He knew the man would return. Ten dollars was a lot. The bounty hunter climbed into his empty truck and drove away.

"We feasted that night," Ella said. "Daddy kill a chicken. We had greens, biscuits, fatback, and punkin' pie. We didn't eat like that except for Christmas." Although Ella didn't know it at the time, her father must have. It would be their last meal together as a family.

That night, Ella's father handed her a burlap sack he used for gathering turnips. In it she placed two picture books, a copy of the *Saturday Evening Post*, her boots, three or four everyday outfits, and the yellow Sunday dress that had been sewn for a cousin but now belonged to Ella. The next morning, while her brothers were still asleep, Ella and her father left in the dark. A neighbor's mule pulled the wagon. The trip from Abita Springs to Carville would take two full days.

On the slow ride west, Ella sat on the front of the wagon. She had never before been allowed to sit alongside her father. Along the way, they stopped to have a picnic under a shade tree. They picked wild blueberries and ate them on the shore of a pond. When they reached the river road near Carville, they parked on the levee and walked down to the Mississippi River. Ella put her feet in the muddy water. Her father suggested she put on her Sunday dress. She changed behind brush at the river's edge. Late in the afternoon, they arrived at the colony gate. A man who appeared to expect them went inside to alert one of the sisters.

"I ain't never seen a nun before," Ella said. "Big, white bird wings on her head scared me stiff." Ella held her bag as she looked at the nun and then back at her father. He nodded and pointed toward the Sister. Ella, in her yellow dress, walked over to the Sister of Charity, who put her arm around her and led her toward the building. They stopped at the door. Ella looked back at her father and waved. From the front seat of the wagon, he nodded again. Then she turned and stepped into the building where she would spend the rest of her life.

CHAPTER 19

One balmy night after the 10:00 P.M. count, Link invited me to his room. He had something to show me. Chains rattled as the guards walked up and down our hallway, yelling that the count had cleared.

"Clark Kent gonna see some shit tonight," Link said, as I walked into his room.

Link's roommate, Bubba, was from New Orleans. He motioned Link to be quiet. We sat for a few minutes waiting for Link's surprise. It was the longest I'd ever seen him silent. Link grinned, like he knew something remarkable was about to happen.

Then, from outside, I heard someone tap on the window. Bubba raised the window, reached out his hand, and helped a man climb into the room. The two men embraced each other.

"You're breaking into prison?!" I whispered.

"Hey," the man said, in a thick New Orleans accent, "what are brothers for?"

Bubba's brother Butch was a free man, but he had complete access to the prison because no guards were stationed at the gate at night. The brothers undressed and exchanged clothes.

Link, who was not very good at whispering, said, "Tell me this ain't some fucked-up shit!"

"Shut up!" Bubba and Butch said in unison. Link put his hand over his mouth.

Bubba climbed out of the prison window, dropped to the ground, and scampered away. Then he would climb under the fence and up the levee, where his girlfriend was waiting with a car. Butch, now wearing his brother's prison clothes, climbed into Bubba's bunk and pulled the covers over his head.

The switch was crazy, but ingenious. Inmates weren't required to wake up for late-night counts. If a guard could see an exposed body part, we were allowed to sleep through the count. The system the Bureau of Prisons employed for identifying escaped inmates seemed reasonable. But the Bureau could never have anticipated this exchange. Tonight, Butch would take a long nap inside our prison and be counted by the guards. His convict brother, after an evening in Baton Rouge, would time his return carefully to walk in the shadows thrown by the ancient oaks and tall buildings to take his rightful place.

"I gotta find a motherfucker to come up in here for me," Link said. Then he looked at me, "You got any black friends?"

Bubba wasn't the first person to escape, briefly, from Carville. The leprosy patients had used the same technique for decades. The patients called it going through "the hole in the fence." Jimmy Harris had gone through the hole to meet family members. Younger male patients left to dance and drink at a honky-tonk in Baton Rouge. The patients spoke freely of their escapes, as if telling old war stories.

One patient, Annie Ruth Simon, and I shared a common past I could never have imagined. "I'm Annie," she said, as she sat with me one day after lunch in the cafeteria.

Annie didn't look like a leprosy victim. If I'd met her on the outside, I might have mistaken her for a librarian or a retired elementary school teacher. She showed no signs of the disease, except for a slight shortening of the nose. Annie Ruth was not much over five feet tall, probably close to seventy-five years old, and not at all self-conscious.

Annie Ruth and her husband, when they were younger, had climbed through the hole in the fence on a fairly regular basis. Their favorite destination was the Roosevelt Hotel in New Orleans. Annie Ruth told me about their exploits in fine restaurants and expensive hotels, so I assumed they came from prominent families. They had no way of earning money at the colony. Annie Ruth laughed about slipping through the fence to attend LSU football games. She and her husband never missed a home game between LSU and Ole Miss.

When I was twelve, my father took me to Baton Rouge to see an Ole Miss/LSU game. I sat in the midst of fifty thousand screaming fans, surrounded by the smell of bourbon and cigar smoke, women in fur coats and men in blazers and topcoats. On the last play of the game, with Ole Miss ahead, LSU was threatening to score. The Ole Miss defense held and we went wild, jumping up and down, slapping hands. But the LSU timekeeper had failed to start the clock on the snap. His error added almost four seconds to the true length of the game. With the extra seconds, LSU's quarterback, Bert Jones, threw a touchdown pass to win the game.

When I asked Annie Ruth if she had been at the game in 1972, her face lit up. "We were there!" she said. "I loved Bert Jones."

Twenty-one years earlier, and twenty-one miles north of Carville, Annie Ruth and I had been in the same stadium. As Annie Ruth and I exchanged memories of the game from 1972, a couple of other patients gathered around. They were all LSU fans. Like all LSU fans, they remembered the historic Billy Cannon punt return on Halloween night in 1959. And like all Ole Miss fans, I brought up the 21–0 revenge Ole Miss took upon the Tigers in the Sugar Bowl to capture the national title.

I felt like an insider as I sat around the cafeteria table with a half-dozen leprosy patients. We told our stories. I was more than an undercover journalist. I was more than an eyewitness. I was participating in a new kind of community. Prisoners and leprosy patients might have been considered outcasts by most of the world, but we were stuck here together. I was still a bit apprehensive about touching them, but I realized they wouldn't want me handling their finances either.

Ella rolled by. "Boy, you always meddlin'." That's what Ella called it when I interviewed people—*meddlin'*. She shook her head, and I went back to making notes.

Jimmy Harris interrupted. "I'm writing a book," he blurted out, even though we had all heard about it repeatedly. "Already got a title. *King of the Microbes*—catchy, don't you think?"

The other patients were looking toward the door behind me. I turned to see a guard walking toward our table.

"Inmate!" he yelled at me. "Get over here."

I had never seen this guard before. He wore a different kind of uniform. He stood tall in the midst of the wheelchairs and walkers. As I walked toward the prison cafeteria, he said under his breath, "I'm gonna write you up."

From behind the latticework dividing the cafeterias, I listened to him chastise the patients.

"You shouldn't talk to inmates," he said.

"You can't tell us what to do," one of the patients said.

"No, I can't," the man said, "but I'm warning you; these aren't choirboys. They're convicts. They steal. They lie. They're dangerous."

"Everybody makes mistakes," Harry said. Harry was shy and usually didn't say much. I was honored that he would stand up for our right to be friends, especially after I had not taken his hand.

As the guard walked toward the inmate cafeteria, Ella called out, "They just childrens."

In the office, five guards peppered me with questions. *You want to go to the hole?! You think your little comments on the board are clever, don't you!? How many times have I told you—you can't fraternize with patients!* The four guards deferred to the new guy. One of them called him "Lieutenant."

I stood quietly until they ran out of things to say. "May I be candid?" I asked.

"Hell, no!" one of the guards yelled.

The lieutenant sat down behind the desk. "Say what you gotta say."

"The patients talk to me," I said. "They're nice. And old. I don't want to be rude. I'm just trying to be kind." I failed to mention my

secret, undercover identity. Then I said, "If you don't want us to fraternize, why did you put us here together?"

The guards exchanged concerned looks.

"You won't be together for long," the lieutenant said.

I wanted to ask for an explanation, but if he knew I cared, he
wouldn't offer any information. I stood still and quiet.

"They're going to be removed," he said. "Relocated."

CHAPTER 20

I was appalled that the Bureau of Prisons would force the leprosy patients to leave their home. Even for me, Carville was turning out to be a good place to serve a prison sentence. Though I wasn't able to live at home with Neil and Maggie, most weekends we spent eight hours together playing, laughing, and telling stories, without the distractions of a television or phone or obligations. The weekends when Linda couldn't bring the kids, I had a long list of people willing to make the drive to Carville. My mother would sometimes collect Neil and Maggie and bring them to spend a Saturday in the visiting room. And my father would drop by unexpectedly, just to catch up. John Caridad, a priest from Gulfport, spent a Sunday afternoon with me. A week later, an Episcopal priest from New Orleans dropped by. My friends from church, the Singletarys and the McCrarys, spent an afternoon at the colony, as did Jack Yelnick, my college buddy from Chicago.

I was one of the lucky ones. Many of the men in Carville, like the leprosy patients before them, had been disowned. Brian Kutinly, a friendly young man from Connecticut who mishandled funds at his home health agency, said his father accused him of ruining the family name. He said he never wanted to speak to Brian again.

But I did have a loving family and loyal friends. Life on the inside could have been much worse. I also loved getting back to investigative journalism. My early success with interviews for my prisoners-in-the-leper-colony exposé had the same feel and energy as my first foray into reporting.

When I launched my first newspaper straight out of college, I called it the *Oxford Times*. For 105 years, the town had been served by a rather complacent daily newspaper, the *Oxford Eagle*. It was part of the elite, the establishment. The editor and publisher had no interest in fighting for underdogs or the disadvantaged. In my view, they did nothing to challenge those in power.

Launching a newspaper that practiced real journalism, I thought, would be one step toward fulfilling my mother's prophecy of "doing great things." I imagined I felt the same excitement as Henry Luce when he started *Life* and *Time*. I knew William Randolph Hearst had started out just like me, with nothing more than an idea and energy. And I understood the power that came with owning a media company. I kept a copy of H. L. Mencken's quote tucked away in my desk drawer: "Freedom of the press belongs to those who own the press."

The inaugural editions of the *Oxford Times* were greeted with enthusiasm. The town was in dire need of a new voice, and ours was greeted with praise. We were the first to report on public meetings, expose conflicts of interest among the powerful in the town, and report the proceedings from the criminal courts in Oxford. We worked to find riveting stories, and we reported them in great detail: the Santa Claus who was shoplifting from the mall (concealing the merchandise in his big red bag); the Peeping Tom who argued in open court that he was conducting research for his romance novel; the town drunk who drove away in an ambulance to expedite his inebriated friend's arrival at the emergency room. I was having fun. I felt like I was on an important mission to make Oxford a better place to live for all its citizens. And the applause was intoxicating.

My undercover Carville story got another boost when a new prison librarian started work. Patty Burkett, a pretty, red-haired civilian from Baton Rouge, whipped the library into shape. She organized the books in a more traditional manner, enrolled us in an interlibrary loan program with the state of Louisiana, and even started a library newsletter for inmates. Rumor was that she had recruited some English graduate students to lead a weekly book club. I was excited about Patty's arrival and volunteered to help out.

She called a meeting of inmates interested in working in the library. On my way to this first gathering, I heard a raspy voice call out from one of the adjoining breezeways.

"Hey," he said. "Over here."

It was one of the leprosy patients. He stood in a dimly lit section of the hallway. Smoke from his cigarette floated around his head, and I noticed burn marks between his fingers where cigarettes had scarred his numb hands. I'd seen him in the cafeteria. They called him Smeltzer.

Smeltzer had a head of thick gray hair, slicked back with hair tonic. He wasn't terribly disfigured, but he had trouble with his hands and feet. He wore shoes with big Velcro strips and leaned on a walker.

He motioned for me to come closer, but I didn't want to breathe in the smoke he had just exhaled. He held a small piece of paper in his claw hand. A prescription. He held it up for me to see.

"Ten dollars," he said.

"No, thanks."

"It feels good," he said in a low conspiratorial voice.

I couldn't believe a leprosy patient was trying to sell me drugs. I told him politely that I didn't want anything to do with drugs, and reminded him we inmates were routinely tested for narcotics.

"No, no," he said, shaking his head in frustration. "A pedicure . . . it's a script for a pedicure. At the foot clinic."

For a moment, I considered it. But I had seen the patients coming out of the foot clinic, legs extended from the wheelchair, toes gnarled and twisted, some with no toes at all. Smeltzer held out the paper, enticingly.

"No thanks," I said.

As I started to walk away, Smeltzer said, "Tell your friends I'm here."

I wanted to ask him why he needed money. The government provided for all his needs. I looked back to see Smeltzer leaning against his walker, prescription held tight in his one hand, a cigarette dangling from the other.

CHAPTER 21

I spent the late summer afternoons walking the inmate track. It was the one place I found solitude. By early afternoon, with temperatures near a hundred degrees, almost all the other inmates, as well as the guards, sought refuge inside, where the thermostats were set on sixty-eight degrees. Only the Mexicans were outside at this time of day. They played handball, but they rarely spoke to me.

The track was a wide concrete sidewalk that circled the perimeter of the prison courtyard. As I walked around and around with no guards watching and no inmates screaming or slamming dominoes on tabletops, I could let down my guard and relax. My imagination ran wild.

I imagined Linda and the kids on the beach at the Singletarys' Florida condominium, running from the waves. I dreamed of adventures with Neil and Maggie. I fantasized about magazines I would launch. I ran through dialogue and scenes, imaginary encounters with people on the outside.

At times, when I couldn't fend off the guilt about the burden this incarceration caused my family, I let anger engulf me. And I walked. I walked fast until I was sweating profusely, sometimes to the point of exhaustion. Awash in self-pity, I thought about all our family would miss together—Maggie's ballet recital, Little Neil's first day of school, Saturdays at Audubon Park, Sunday brunch at La Madeleine. And, worst of all, Neil's and Maggie's birthdays.

I didn't know what to do, so I walked and daydreamed. Hours disappeared as the rhythm of walking pushed me into a state of meditation. Days slipped away. I realized I could fast-forward time, one repetitive circle after another. But it wasn't my first experience with manipulating time.

My initial success as a journalist in Oxford roused the established daily paper from its slumber. The publisher of the established daily, the *Eagle,* launched an aggressive marketing campaign depicting us as troublemaking outsiders. We lost some advertisers. I believed my commitment to good journalism would prevail, but soon I ran short on cash.

My biggest advertiser was a local grocery store. On the last day of March 1986, the owner had promised to pay a bill—for almost $5,000. Taking the grocer at his word, I wrote payroll checks. The check from the grocery store owner did not arrive.

With two newspapers in town, customers could easily move advertising from one paper to the other. As desperate as I was for that check, I was equally desperate to keep the grocer as a long-term customer, so I didn't want to be too pushy and risk alienating my biggest client.

The next morning I called the grocery store. The owner was out of town and would be gone for two days. No checks were to be issued in his absence. I put the telephone down and looked at my clock: 9:00 A.M. At 10:00 A.M. the bank would stamp my payroll checks, "Insufficient funds." I had one hour to raise $3,000.

I had $400 in my personal account at another bank in town. It occurred to me that I didn't really need $3,000. I needed time. I needed two days. Sixteen working hours. Just enough time for the grocer to return and deliver the promised funds.

I opened my personal checkbook and, despite having a balance of $400, wrote a $3,000 check to myself, from myself. On the bottom of the check I wrote, "loan." The following day this check would bounce without some kind of coverage, so the next morning, a Thursday, I wrote a check from my business account to my personal account for $3,000. On the subject line I wrote "repayment."

The next day, the grocer returned. I drove to his store and picked up a check for $4,700. I deposited the funds in my business account to cover the check I'd written the day before.

No one seemed to notice the transfers. All was well once the grocer fulfilled his promise. I discovered that with two checking accounts at different banks I could do something I'd never dreamed possible. I could buy time. And no one would ever know.

Walking in circles around the track, I understood a gerbil's drive to run inside a wheel. I felt that same instinct. I walked so many hours, the concrete wore down the soles of my boots. By late August, I had lost almost thirty pounds, leaving the fat behind me, one drop of sweat at a time.

On my laps around the path, I passed Ella in her antique wheelchair, navigating the walkways. I waved at Harry and Father Reynolds riding their old-fashioned bikes. Jimmy and Sister Margie nodded as they trekked by on their three-wheelers. They were going in circles, too. And I wondered if it gave them some sense of escape or helped them pass the time.

One particularly hot day, I noticed Ella making more than a dozen trips from her room to the patient canteen and back. I wanted to ask if she had heard about the plans to relocate the patients, but I didn't want to upset her.

As I turned the corner by the breezeway that ran between the prison and the leprosy sides, I heard her call out for help. I looked around for a guard. Through the screen, I saw the handles on her wheelchair, but I couldn't see her. She called out again. I yelled that I'd be right there. Ella had slid down in her chair. Too exhausted to pull herself up, she was about to fall onto the concrete floor. I reached over the back of her chair, put my hands under her arms and gently pulled her into a sitting position.

"You OK?" I asked.

"Tired," Ella said, her voice weak.

The temperature, even in the shade of the hallway, was in the nineties. I was about to tell her to be more careful, to stay in the air-conditioning, when she asked if I would push her to her room. I looked around the inmate courtyard again. No guards. No one else to help. I pushed the old chair around to her dorm. The faster I pushed, the faster the hand cranks rotated. I slowed down after Ella's head listed to the left. I worried the spinning handles might hit her.

I held open the door of her dorm and waited for her to take over.

"To my room," she said, nodding down the hall.

"I'm not supposed to go inside," I said.

"Will they put you in jail?" she asked, exhausted, but with a smile.

She had a point. I figured I couldn't get in too much trouble for aiding an eighty-year-old woman with no legs. As I pushed Ella into the cool hallway of her dormitory, I realized that for the first time I had touched someone with leprosy. I made a note to pay attention not to touch my own face until I got back to my dorm to wash my hands.

No one saw me push Ella into her room. No guards. No other patients. No inmates. No one knew where I had been.

CHAPTER 22

On one of my afternoon walks, when the shade from the trees started to spill over onto the concrete track, Link joined me as I made laps around the inmate courtyard.

"Why you hang out with them leopards?" Link asked. "You gonna catch that shit!"

"I thought you wanted to catch it," I said.

"I do, but I gotta figure out how to make some cash 'til I get that shit." Link looked at me and smiled. "Teach me how you stole money from them banks!"

"What kind of relationship do you have with your banker?" I asked.

"Fuck," Link said, "I ain't never been inside a motherfuckin' bank!"

I explained to Link that kiting checks required patience, a unique relationship with financial institutions, and was dependent on the appearance of stability. And I reminded him that I didn't actually get any money for myself. "It's really about buying time." I told him. Link didn't follow. "It's complicated," I said, hoping he'd lose interest.

"My shit complicated, too!" Link said. He launched into the difficulty of supply and demand in the drug business. Link had been forced to adapt. When the drug money dried up, he said, he fell back on carjacking.

"You didn't."

"Fuck, yeah, I did!" he bragged, adding that carjacking took patience, too. "I hide in the bushes at the Popeyes drive-though," Link said. "I wait there until a woman come through in a Mercedes or a BMW. Late model," he said, "Couldn't be no man in the car." Sometimes Link waited for hours. "Then after she get her food, I jump out, point the gun at her face, and say, 'Get out the car, bitch!'" Link threw his head back and laughed. "Them ladies would jump out the car—leave their keys, their purse, and their chicken. So I had something to eat on the way to the West Bank."

"You planned your carjackings around chicken!?"

"You ever had Popeyes chicken?!"

I thought about Neil and Maggie and Linda. Driving through a Popeyes in our vintage Mercedes. "You never did that with children in the car?" I asked.

"What the fuck difference do that make!?"

"That sort of thing affects people. They have nightmares. They are afraid to leave their houses. They get depressed. They live in fear after that sort of experience."

"You think I give a fuck!?" Then Link said, "Why you always worried 'bout other people?"

I asked Link to imagine what it would feel like if someone carjacked his mother.

"My momma ain't got no good car."

I obviously wasn't getting through. "Did you ever get caught?" I asked.

Link was almost caught once, but only because he abandoned his well-laid plan. He made a spontaneous decision.

"There was this Little Debbie truck," he said, "making stops on St. Charles Avenue." Link said the driver left the engine running to make a quick delivery. Link assumed the money bag was inside the truck. "I jump in, take off. That motherfucker was a stick shift!" Link jerked his body back and forth as he described the drive down St. Charles Avenue. "Zebra Cakes flying all over the motherfuckin' place! Drove five blocks. Then hit a motherfuckin' streetcar pole!"

"Were you hurt?"

"Fuck, yeah," Link said, "racked my nuts."

Link's stories were outrageous, but he had a sharp sense of humor, and I sensed, deep down, he was a good person. I didn't understand how he could randomly point guns at people.

"If you saw me walking down the street in New Orleans, would you point a gun at me and take my money?"

"No," Link said. "I know you."

"Then how can you just point a gun in a woman's face and never think about it again?"

"When you on crack," Link said, "you don't see no faces."

Link joined me often. Most of his youth had been spent in and out of juvenile detention centers. And Carville was his sixth go-round in prison as an adult. Link told me he was only twenty-seven years old. He had been shot four times. "Bullet burns like a motherfucker," he said. "How many times Clark Kent been shot?" he asked, laughing at the absurdity of the thought. "Fuck, I forgot, bullets just bounce off Clark Kent!" Link threw his head back, laughed, and walked away.

Link and I were different in many ways. But I had trouble in my twenty-seventh year, too.

On a humid summer morning in 1987 in Oxford, I received a call from my banker, a lovely young woman named Wendy, who had just been promoted from teller to loan officer. When I entered her office, I saw the look on her face, and I knew. My heart pounded fast, but I tried to hide my fear.

"This is the hardest thing I've ever had to say," she said in her sweet southern drawl. Wendy looked down at her desk and said it had come to her attention that I had been kiting checks. Close to tears, she told me she was going to be forced to call my loans. She would have to close my checking account, at least for a while. And, as if the words almost wouldn't come out of her mouth, she was required to complete a form that would notify the FBI of my activity.

I left her office and went to see my only local investor. With no-where else to turn, I explained what had happened. When he asked the amount of the overdraft, I told him: "About $8,000." He seemed relieved. He called a meeting of the other investors. They agreed to invest another $30,000. We opened a new account at another bank in town. The investor's secretary would serve as bookkeeper. The next day, I covered the overdraft, paid off the loan, and went about my business of publishing the *Oxford Times*. I got the sense that, in spite of my poor financial management, my investors liked what I was doing with the newspaper. I had also convinced myself that my actions were justified because the newspaper was important to the community.

I didn't tell Linda about the incident. I was rattled, but I had a newspaper to get out. No time to waste with worry. I would tell her when the FBI called. But they never did. The incident was never investigated. I never heard another word about it.

In a way, I did feel bulletproof. And one thing was clear. As long as I raised the money to pay my overdrafts, nobody seemed to take kiting too seriously.

I also figured if the investors had put up another $30,000 after I'd been caught kiting, they would follow me anywhere. I embarked on an aggressive campaign to attract new advertisers. I fired our delivery boys and shifted to direct mail delivery of the *Oxford Times*. It added $2,000 per week to our expenses, but I was certain it would pay off. I also hired more writers. Better stories, more complete coverage, would help me compete against the *Eagle*.

But the new advertisers never arrived. Seven months later, I stood in a U.S. bankruptcy court in Oxford, Mississippi.

My investors lost $90,000; local vendors lost more than $50,000.

I had failed, but I did not regret trying. And I refused to end my publishing career in failure. Even as I stood in bankruptcy—an attorney at my side and creditors to my back—I knew I would try again. I would take a lesson from the *Oxford Eagle*. Its editorial restraint had a financial upside. I would move to Gulfport, a market thirty times the size of Oxford. I would avoid conflict with those in power. I would publish a glossy magazine that showcased the Mississippi Coast. I would focus more on making money and less on changing the world. I had already talked to Linda about my plan. She agreed to get a job to support our family while I raised money for the new venture. All I needed was $50,000.

As I turned to face those I owed but could not pay, I noticed a singular absence in the courtroom: reporters. The *Eagle*'s indolence extended even to my demise. I had launched the *Oxford Times* precisely because the established daily was unresponsive to situations like this. Now, I was the beneficiary of their lethargy. There would be no detailed account of my debts, no questions about shuffling funds, no printed opinions surrounding my failure, and no testimony from local merchants who had lost money. No one was there to tell the story.

CHAPTER 23

Doc's job as an office clerk afforded him access to inside information. As a clerk, he made copies, transported memorandums between departments, and overheard conversations between guards. Sometimes he made extra copies of memos to pass among the inmates. We often knew about prison initiatives before the guards did.

I asked Doc if he'd heard that the patients would be relocated.

Doc had read all about the plans to move the leprosy patients. One plan called for handing out an annual stipend so patients could pay for their own housing. Another plan required displaced patients to be relocated at a nursing home. Another option included construction of a new set of dorms on the other side of the colony so the inmates and patients could not commingle.

The Bureau of Prisons planned to turn Carville into a massive prison with more than a thousand inmates. And the primary objective of each proposal was simple: remove the leprosy patients.

On a rainy Sunday afternoon, when Doc took a break from his reading, I told him about my plans to write a magazine exposé about Carville. I recounted Ella's journey to Carville. I told him about my shared history with Annie Ruth at the LSU football games. I gave him a detailed account of Link's carjacking escapades.

Doc raised an eyebrow. "Don't believe everything these people tell you."

"The inmates?" I asked. "Or the leprosy patients?"

"Both," he said. "They make this crap up." Doc, in a rather condescending tone, reminded me that the patients were institutionalized. The inmates, he said, had nothing better to do with their time than tell lies.

I felt like I had disappointed him. Like a younger sibling who was naive and gullible. Doc was annoyed that he had to live with these people. They were beneath his intellect. He didn't want them as friends. But I believed the stories. The details were too vivid to be fabricated. My instincts, everything I had learned as a reporter, told me they were not lying.

"They don't have any reason to lie," I said.

Doc didn't answer. He went back to reading his medical journal. A moment later, he said, "Just don't believe everybody."

The photograph that appeared in Entrepreneur *magazine, 1991.*

CHAPTER 24

Carville was full of men whose grand schemes trumped common sense. Steve Read and I were two of the best examples. Three weeks after I declared bankruptcy in Oxford, my family and I moved to Gulfport, Mississippi. We lived rent free in my aunt Viola's empty house. Linda took a job as a receptionist at a local advertising agency. I stayed at home, wrote a business plan, and sought out investors. My mother and father encouraged me to get a job. "You're obviously not cut out for business," my father told me. But he didn't understand. I'd read stories about businessmen who rebounded from bankruptcy. "Success is always built upon failure," I'd read. I believed in second chances, and I was perfectly positioned for a comeback. Because my financial troubles in Oxford never hit the press, my reputation in my hometown of Gulfport remained unblemished.

I presented investors with my business plan for a glossy lifestyle publication for the Mississippi Gulf Coast, a metropolitan region situated between New Orleans and Mobile. Within three months, I had convinced a local businessman to invest $50,000 and I launched *Coast Magazine*.

In the first year of publication, our circulation grew to twenty thousand. It surpassed our wildest expectations. Linda quit her job at the advertising agency and joined me as editor of the publication. Our magazine made the people and places of the coast look beautiful. New investors approached me about getting in on the action. I launched a sister publication, *Coast Business Journal*. Soon, I bought out the businessman who initially invested.

I hired a ragtag crew in the early days, men and women with real talent who had never been given a chance. I trained them and, for the most part, they rose to the occasion. An elementary school teacher's assistant ascended to the position of editor. A recovering drug addict and ex-con handled door-to-door subscription sales. A high school dropout worked his way up to the position of art director. A marginally literate woman became a top salesperson among the restaurateurs and nightclub owners of Biloxi.

I owned a burgeoning media empire, and I was having a great time again. The advertising dollars rolled in, as did the awards. The plaques and trophies and certificates covered the walls of our offices. Linda and I moved into a new home one block from the beach. I bought a vintage Mercedes for Linda, a boat for myself.

Our magazine featured never-before-published photos of Elvis Presley, as well as rare images of Jayne Mansfield taken the day before her fatal automobile accident (the starlet was photographed straddling a large cannon on an island just off the Mississippi coast). Each issue featured exquisite architectural photography highlighting the finest homes in the region, including full-color spreads of the home of Judge and Mrs. Walter Gex. We added fun departments like "People to Watch," "20 Questions," and "Nostalgia Yearbook," wherein we scanned high school yearbook photos of bank presidents, mayors, beauty queens, and politicians, along with a few notorious residents serving prison time.

I set out to build an empire and everything that went along with it. As we celebrated the first anniversary of *Coast Magazine*, I presented a five-year plan of ascent to my employees.

The first year, I would buy *Louisiana Life* magazine, a financially distressed publication with a devout subscriber base that in its heyday had won a National Magazine Award. It would be the first acquisition in building a network of city and regional magazines.

The second year, I would purchase real estate—the Hewes Building, a five-story structure across the street from Hancock Bank. Equipped with an antique, hand-operated cage elevator, replete with a full-time operator, the building would serve as our corporate head-quarters.

In year three, I would purchase a yacht. I would christen it *Magazine*, and it would be available to all employees to entertain clients, advertisers, and prospective investors.

In year four, I would launch a weekly trade journal entitled *Business South*. The weekly tabloid would be sent free to every business owner in the southeastern United States. It would function as a quasi *Reader's Digest* for businesspeople in the South.

And in year five, in a crowning triumph, I would purchase the *Sun Herald*, the coast's daily newspaper with gross receipts in the $20 million range.

The community embraced our mission because we made the coast look good. During an introduction to a local Rotary Club, an economic development expert called us a "shining star" in an otherwise dismal economy. My mother and father beamed with pride when their friends complimented my good work. And my friends seemed delighted with our achievements. My employees felt the same way. They were excited about the prospects of our future, too. And I promised, if they worked hard to reach my goals, they would all be rewarded handsomely.

In less than a year after bankruptcy, I was back, and on my way to building a media dynasty.

But even my entrepreneurial spirit was overshadowed by Steve Read's. As CEO of Read Industries, at age thirty-five, Read had owned a fleet of private jets chartered by the likes of Reba McIntire and golf pro Jack Nicklaus. Oil Mop, a corporation he bought out of bankruptcy in the late 1980s, was the first to arrive at the *Exxon Valdez* spill in Alaska, resulting in a multimillion-dollar contract for Steve. He expanded his regional airline business, L'Express Airlines, just before jet fuel prices skyrocketed.

Steve had been sentenced to four years for money laundering, but he was able to accomplish one thing most of my fellow inmates could not. He retained his assets. His wife drove a Range Rover, lived in an opulent New Orleans condo, and wore Ralph Lauren ensembles.

Steve used his money to hire an entourage of inmates to work as his personal employees. He had an inmate chef who prepared meals so he could dine privately. A former personal trainer guided him through a weight-lifting regimen and consulted with his chef about Steve's dietary needs. A maid washed and ironed his clothes, made up his bed, swept and mopped his floor, and provided other domestic services. He had his own inmate barber. His affluence, at least compared with the other inmates, was apparent to everyone. Link called him "Richie Rich."

Steve had a big smile, perfect teeth, and a year-round tan. He sunbathed on the shuffleboard courts in the inmate yard. A reporter for the *Baton Rouge Advocate* who visited the facility saw Steve and his inner circle soaking in rays. The reporter described the scene as convicts basking in the sun like repentant lizards. After the story, Steve picked up on the term. "Off to repent," he would say, gathering his towel and baby oil.

Steve wasn't very popular with the inmates outside his employment circle. With his entourage following, he publicly taunted prisoners about their crimes. To Vic, a man charged with arson, Steve would yell, "Hey, Vic, got a match?" To Semmes, a car dealer who rolled back odometers, he called out across the prison grounds, "Semmes, can you do a little work on my release date?!" And to Daniel Stephens, the banker from Texas who had his Thoroughbreds assassinated, Steve would burst into a loud horse's neigh.

But Steve took a liking to me. We'd never met prior to Carville, but we had a business relationship. I published *Louisiana Life* magazine, which served as the official, in-flight magazine of Steve's regional airline. In return, his company purchased the back cover advertisement in our magazine at a cost of $5,000 per issue.

Steve asked me to join him in his room on weekends to share his specially prepared meals. And he promised, as soon as a slot opened up, to include me in his invitation-only Friday game night where former executives—men who once wielded great power, but now were not allowed to possess paper money—gathered to vie for economic domination in a game of Monopoly.

CHAPTER 25

Steve, the ultimate entrepreneur, managed to get the best job in prison. He was clerk to the chaplains. No guards supervised him. He answered only to Father Reynolds and Reverend Ray, the Protestant minister. He also organized the call-out lists that allowed inmates to leave the prison side to attend church, prepare for services, practice for choir, polish brass, organize hymnals, or perform other tasks to keep the chapels in good shape.

Steve had access to the chaplains' library. He pulled a few strings to get permission for me to assist in organizing the collection. Steve and I spent hours poring over the books and videos and collections. I had access to documents the prison library could never acquire, including historic films and texts about the colony.

I discovered that Carville was planning a one hundredth anniversary celebration. Nearly a century had passed since 1894, when the first victims of leprosy arrived late in the night at the abandoned plantation. The anniversary event would include a barbecue, a golf tournament, speeches, and a visit from Louisiana's reigning governor, Edwin Edwards.

I came across some fascinating facts in Father Reynolds's study. One book speculated that St. Francis of Assisi had not experienced stigmata, but actually suffered from leprosy. The open wounds from advancing leprosy on his palms and forehead were, according to one historian, mistaken for the wounds of Christ.

Another book, about a leper colony in Hawaii, especially piqued my interest. The settlement, established before Carville was the national leprosarium, was situated at the base of stunning cliffs in the Hawaiian Islands. In the 1860s, leprosy spread through the population. Hawaiians had little immunity to the disease, thought to be carried by Chinese immigrants, and by the 1870s more than eight hundred men, women, and children were quarantined.

Father Damien, a priest from Belgium, served as the resident priest to the lepers of Hawaii. Not much was known about leprosy in the 1870s, and most people, even physicians and priests, were terrified of the disease. Damien conquered his own fears, subjecting himself to repeated, extended exposure to the infected. He invited victims of the disease into his cottage. He embraced sufferers who had open wounds. He entered the isolated death huts to deliver last rites.

When he arrived on the island in 1873, he referred to those confined to the settlement—and himself—as "We lepers." It was foreshadowing. Within the decade, Damien would contract leprosy. And when he did succumb to the disease in 1887, in the eyes of many, he left this life as a saint. A nun who sat at Damien's deathbed claimed that, in passing, his face had lost its disfigurement.

The biggest perk that came along with volunteering in the chaplains' library was freedom of movement. The library was the farthest point away from the prison side of the colony. As I walked to and from work each afternoon, I watched the Sisters of Charity, baskets in hand, wandering under the trees in search of fallen pecans. I noticed the smile and soft touch they shared whenever they passed a leprosy patient.

The sisters' kindness inspired others, too. Carville's top official, Dr. Jacobson, erect and imposing in his white naval uniform, softened when he greeted a patient. He treated each patient with tenderness and respect.

Many days I would pass Harry while he was pushing another patient to the foot clinic, or Ella on her way to the canteen, or Father Reynolds riding his bike, his brown Franciscan robe dragging just above the concrete floor.

Almost every day as I walked to Father Reynolds's study, I passed Sister Teresa Pazos. She was one of the Sisters of Charity, but she was also a leprosy patient. The disease had ravaged her nose and fingers. She navigated the corridors in a huge electric wheelchair. She could not have weighed more than ninety pounds. I had no idea if she contracted leprosy while caring for patients here, or if she chose to come to Carville after her infection. The soft hum of her wheelchair motor foreshadowed her appearance from around a corner. She never made eye contact with me, but I looked forward to seeing her. Whenever we passed, a visceral warmth flowed through my body. Just being in her presence made me feel light and peaceful.

I didn't fully understand why she had that effect on me, but I was beginning to feel certain about one thing. Carville was sacred space.

One late summer afternoon while I was dusting the tops of books, Father Reynolds passed through his study, and I asked how long he had been at Carville.

"I came here in 1983."

He was approaching a decade. The same amount of time it took Damien to contract the disease in Hawaii. Father Reynolds's willingness to sacrifice so much, to confine himself here with the very last leprosy patients in America, made me feel like I was in the presence of a righteous and devout saint.

I told Father Reynolds that I admired his bravery. He looked a little confused. "You came here to help the leprosy patients," I said, stating the obvious.

Father Reynolds smiled. "The golf course was the big attraction."

I knew he was downplaying his own work. Father Reynolds was the first priest to leave the altar during the Eucharist and greet the patients. He touched their hands. He welcomed newcomers with a hug. He was committed to bringing faith and compassion to the outcasts.

"Do you worry that you'll catch it?" I asked.

Father Reynolds's perpetual smile left him. "No, of course not."

Since the time Carville had opened a hundred years ago, he explained, no health worker had ever contracted the disease. I wanted to ask about the little nun zipping about the colony, but I didn't.

"No one has been forced to come, or stay, here in over thirty years," he said.

Until that moment, I had believed the patients were still imprisoned. I couldn't imagine why anyone would choose to stay inside the colony walls.

Then Father Reynolds told me a story. In the late 1950s, after medications were developed to control the spread of leprosy, the gates of Carville were opened. At that time, 297 patients lived at the leprosarium. One year later, 281 remained inside. Ella, Harry, and others, who were brought here involuntarily—sometimes in shackles—chose to stay, even after they had been set free. For them, freedom was more terrifying than imprisonment. The stigma of being labeled a "leper" had cut as deeply as any physical scar.

PART III

Fall

CHAPTER 26

Frank Ragano, Jimmy Hoffa's lawyer, was terrified he would catch leprosy. He refused to touch doorknobs or handrails, compulsively wiped the library typewriter keys, and scrubbed the cafeteria's plastic utensils with his shirttail. So, naturally, the guards gave him a job picking up cigarette butts outside the cafeteria.

For five hours a day, Frank picked up the trash and cigarette butts left by the 500 inmates, 50 guards, and 130 leprosy patients who passed by. He wore sterile rubber gloves and a paper mask.

Link immediately picked up on Frank's idiosyncrasy and one afternoon, while standing in line for dinner, yelled, "Who that is? Motherfuckin' Howard Hughes!" A leprosy patient in a motorized wheelchair stopped to listen to the commotion.

"Look at that motherfucker," Link announced, "wearin' a goddamn mask and gloves. This ain't no fuckin' laboratory!" Link grabbed the stub of a cigarette from one of his friends and flicked it past Frank. It bounced against the wall. Link laughed again, knowing that Frank would have to pick it up, and the inmates standing in line laughed, too. Frank did not move.

The skin on Frank's arms was thin and spotted with purple patches where capillaries had ruptured. The tiny wrinkles on his forearms looked like small waves on a lake. This man, who was now being ridiculed by a drug dealer, in front of leprosy patients, had been the highest-paid criminal defense attorney in Florida. He had argued before the U.S. Supreme Court and convinced them to overturn his clients' criminal convictions. He flew into Cuba in 1959 and stared down Castro's lieutenants to get Santo Trafficante released from custody. For all his courtroom brilliance, Frank had no skill for dealing with this kind of humiliation. He stared straight ahead in his flimsy paper mask, arms at his side.

Later that afternoon, I saw Frank sitting on a bench in the inmate courtyard. I'd heard he was writing a book about his days as a mob lawyer, so I asked him about Jimmy Hoffa.

"What was he like?" I asked.

"Jimmy always preached 'Charge a pistol. Run from a rifle,'" Frank said, grinning. During one of Hoffa's criminal trials, a man pointed a pistol at his head. Hoffa charged like a bull toward the assailant and wrestled away the gun.

"What's one thing," I asked, "about Jimmy Hoffa that no one else knows?"

Frank looked surprised. "That's the first question my book editor asked." He leaned in a little closer, as if to prevent anyone from hearing his words. Frank had successfully represented Hoffa for years. During trials they had worked late into the night, exchanging confidential information. I was about to hear a private revelation about one of the most infamous men of the twentieth century—something no one else knew about the notorious Jimmy Hoffa. Everything was coming together again for me as a journalist.

Frank knew how to make the most of a dramatic moment. He pressed his lips together and swallowed, contemplating whether to reveal his secret. Finally, he motioned for me to move closer. Then he whispered, "Hoffa loved to fart."

Frank smiled and reiterated, "He loved it."

I imagined Hoffa banging his hand against the Teamsters podium, red faced, veins protruding from his neck, relieving himself during a sudden burst of applause. Or sticking his index finger in the face of Attorney General Robert Kennedy and barking, "Hey, Bobby, pull this." Or leaving a little something for the passengers on the elevator at his headquarters in Chicago.

I laughed and said, "That's not going in your book, is it?"

Frank shook his head and smiled again. The secret was all mine. But Frank obviously had other secrets. "Do you really know who killed Jimmy Hoffa and JFK?" I asked.

"Yes," he said. "Yes, I do."

Little Neil, Maggie, and me on a fall visit.

CHAPTER 27

I missed my cologne. For years, I would douse myself with British Sterling every morning. I even kept a little bottle in my office, for a refreshing boost during the day, hoping the people around me would associate my presence with a classic, soothing smell.

I hated to walk into the visiting room to see Linda without my fragrance. In prison, cologne was contraband. My signature scent was beyond my reach, but I soon discovered a source. On Monday afternoons, the magazine subscriptions arrived in the prison library. I volunteered to help put them in their proper order so I had first shot at the scented advertisements bound into *GQ*, *Esquire*, and other fashion titles. As I organized the periodicals and flipped through the pages, I tore out the aromatic inserts. I stockpiled the samples in my locker, and before visiting hours, I opened the strips and rubbed them against my shirt until the fragrance permeated my green uniform. For the rest of the day, whether inside or out, cool or perspiring, I was the best-smelling inmate at Carville.

But the allure of cologne was no match for reality. Things weren't going so well with Linda. The financial arrangements I had made to support my family while I was in prison were falling apart. As was our relationship. Lately, Linda and I had been arguing—on the telephone and in the visiting room. During a recent visit, Maggie told us, "Y'all don't fight. Y'all hug." But a hug wouldn't solve the money problems.

I had loaned money to a friend who said he would send a monthly check to Linda, but his lawyer advised him to stop making payments when we filed bankruptcy. It was money Linda had counted on, and I had no way to persuade my friend to do otherwise. Prior to reporting to Carville, I had also sold advertisements for a publication. Two of the advertisers were to send payments to Linda, but they had failed to do so. Linda was going to have to go back to work, and the kids would go to after-school care, something we'd hoped to avoid during this year while I was in prison.

To make matters worse, she ran into Bill Metcalf, the owner of the region's premier publishing company, a generous man who made me publisher of *New Orleans* magazine while I was being investigated. I had left many unfinished projects when I went to prison. One project left prepaid contracts that couldn't be fulfilled without more sales. Bill's frustration with me was justified; I'd left him with a mess. But when Bill let Linda see his anger, she was again reminded of the terrible situation in which I had left not only my family but also my friends.

After seven years of marriage, we sat in a prison visiting room and argued. When we married, I assured Linda that, together, we would be free and independent. That we would have extraordinary adventures. I had not kept my promise. Now we were living off the charity of others. My mother was paying the rent on Linda's apartment in the French Quarter. Grandparents paid tuition for Neil's school at Trinity Episcopal; Linda's parents were paying for Maggie to continue at the Louise McGee School, a private school just around the corner from the Garden District homes of Anne Rice and Archie Manning.

About the only thing Linda hated more than charity was pity, and she was getting it from everyone around her. From well-meaning friends. From fellow church members. Even from family.

And the apartment in New Orleans was a problem. It was located one floor above my mother's weekend apartment. Mom thought it would be a good idea to be close enough to help care for Neil and Maggie. She was right, except that, despite my incarceration, she still talked about the high hopes she had for my future. For Neil and Maggie, it might have been a good thing. For Linda, who was trying to make a clear, objective decision about what to do with her life, my mother's praise of her precious son was hard to swallow.

On a visit in early fall, when Linda and the kids were planning for the beginning of school, I thought it was time to talk to Neil and Maggie about Daddy's camp. If the talk of jail and prison conjured dangerous images, like the psychologist had said, I assumed it wouldn't affect them now, since they had seen the place. They actually seemed to have a good time here. They loved playing with the other inmates' kids; Neil particularly liked the vending machines that held frozen pizzas, Hot Pockets, and ice cream.

They could see that there were no bars, no really dangerous people. And I didn't want them to be caught off guard at school if someone brought up the fact that their father was in jail.

"I want to talk to you about camp," I said.

"I forgot," Maggie said, "why did you have to come to camp?"

I reminded Maggie of what we had discussed—that I got a little bit greedy, wanted to buy too many magazines, and used money I wasn't supposed to use to buy them.

"Boy, Daddy," Maggie said, "you must've really loved magazines."

Then I told them that this place, Carville, wasn't exactly a camp. It was a prison camp. Kind of like a jail, but different.

Maggie had taken a bite of cookie. She looked up and asked, "Are you a bad guy, Daddy?"

"Of course not," I told her. "Remember, camp is kind of like time-out for adults. Only adult time-out lasts longer than kid time-out."

"Did you used to be a bad guy?" Maggie asked. I shook my head. I hated that I had put my daughter in a situation where she would have to ask such a question. I looked over at Little Neil. He didn't ask any questions. He stared down at the floor. Maybe he didn't want to talk about this. Or maybe he just wanted to go outside and throw the ball.

I took Neil and Maggie out to the visiting room deck and gave a coupon to Bean, the inmate photographer. Prisoners could purchase a family photograph for one dollar. All proceeds went to the inmates-in-need fund. Linda said she didn't want to be in the picture, so I put

Neil on my shoulders, and Maggie stood on the bench next to me. As we posed for the photograph, Steve Read walked by.

"Ah," Steve said under his breath, "prison memories." Neil and Maggie didn't seem to notice, so I let it go.

I continued to write letters to Linda and the kids every night, and I made collect calls from the pay phones that lined the walls in the hallways. But some nights the phone line was too long, and I went to sleep without speaking to them. Other nights, I would call about 9:00 P.M. to say good night to Neil and Maggie. Sometimes, I would call late at night so Linda and I could have a conversation without the kids overhearing, though it wasn't really private since inmates were usually standing close by waiting for their turn.

Every three minutes, a recording interrupted the telephone conversations: *This call originated from a federal correctional facility.* It was designed to discourage phone scams, but when it interrupted my conversations with Linda, it was just another reminder of where we stood.

The night before Neil started first grade, I called to talk to him. It seemed like an inopportune time. Linda was cooking dinner; Maggie was running around the house screaming; Neil was doing something that clearly irritated Linda. They had been arguing.

I wished I could have been there to help. I would have taken the kids into the courtyard to play, or I would have offered to cook dinner to give Linda a break. I wanted to climb through the telephone line, hug the kids, and tell Neil his first day of school was going to go just perfectly.

Linda handed the phone to Little Neil, and he took it into the closet and shut the door. He was crying. He didn't cry very often. I told him how much I loved him, how proud I was of all that he did, but it was no substitute for a hug. If I'd been there, I would have picked him up, put his head on my shoulder, rubbed his back, and rocked with him until he felt better.

The best I could do was listen. "Tell me what's going on?" Neil tried to keep from crying, but he couldn't speak. "Tell me what you want, buddy?"

He cried until he caught his breath, and then the words came out. "I want you, Daddy."

CHAPTER 28

Initially, I couldn't fathom why the federal government would decide to put inmates in the same facility as leprosy patients. Link and his buddies claimed we all were part of a secret government experiment. He was certain we would catch the disease, which, of course, was fine with him since he planned to cash in on the situation. Link's theory was corroborated by other inmates, who pointed to the experiments on prisoners at Tuskegee and in Nazi Germany.

The logic behind this experiment was hard to follow. Did some bureaucrat announce, "Hey, I've got a good idea. Let's put federal convicts in with the leprosy patients!"?

But now I was beginning to realize what an insult it was to the leprosy patients. Despite how the inmates felt about it, for the patients, it was another slap in the face. That the federal government thought nothing of moving criminals into their home said a lot about their standing.

When Link was finally given a job, he was assigned to the cafeteria, an assignment that brought him into contact with the patients. He intentionally bungled his tasks. He let pots and pans pile up in the wash room, he filled the salt shakers with sugar, and he ran the floor buffer over its own cord and burned up the motor. The guards wisely decided not to let him get near the food.

Link wasn't happy about having a job, and his opinion of Carville was changing.

"This place is all fucked up!" he said.

"It's not so bad."

"It is. Ghosts everywhere. About a thousand of them leopards died in this motherfuckin' place. And the goddamn Mississippi River flows north!"

I'd heard other men say the river flowed north at Carville, but I found it hard to believe.

"Link," I said, "I thought this place was like a country club?"

"It is!" Link said. "A fucked-up country club!"

Though he spent most of his day sleeping in the back of the walk-in cooler, Link was assigned to the patient cafeteria to help pick up trays and dishes after breakfast and lunch. Most days, I helped him. But a guard usually had to rouse him from sleep. One day when I was preparing the menu board, the patients left an unusual mess. Link entered the room and saw the pile of trays left by the patients.

"Goddamn leopards!" he yelled.

I looked up and saw Ella still eating. Harry, who was leaving, stopped and turned around. He looked like someone had just punched him in the stomach. Stan and Sarah, the sweet blind couple, reached across the table and held tight to each other's hands.

I felt terrible that Link had hurt them. They had chosen to stay in Carville, their safe haven, to avoid this kind of pain. But with the arrival of the inmates, the stigma of leprosy had slowly crept back into their home.

At that moment, I saw in their expressions just how vulnerable they were to the odious label. I promised myself I would not view my new friends as "lepers." And I made a commitment not to use the word that had caused them so much suffering.

I called Link over to the menu board. For all of Link's street savvy, he wasn't so good in the sensitivity realm.

"They hate that word," I whispered. "Plus, you're saying it wrong."

Link shrugged. I led him over to the corner and talked in a low voice. "Link, a leopard is large cat." I looked at him to make sure he was following. "People with leprosy are sometimes called lepers—but they really, really hate it."

"What the fuck?"

"They hate being called that as much as certain people hate being called certain words," I said, hoping Link would make the connection.

"Nigger!" Link yelled. "You mean nigger?"

"Well," I said quietly, hoping Link would follow my lead, "that's one of them."

"I say that word all the motherfuckin' time!" Link said.

I was about to launch into an explanation about the subtle difference in delivery, source, history, perspective, but instead I just asked Link again not to call them lepers—or leopards.

"What the fuck you want me to call them?"

I didn't know. But I would find out.

Link and I finished with the trays; he went back to the cooler for a nap, and I apologized to Ella for what Link had said.

"It don't bother me none," she said. "That word in the Bible. And there ain't no leprosy in heaven."

Ella's attitude was amazing.

"We do our sufferin' down here," Ella said. "Jesus gonna be waitin'. So it don't bother me none." Ella was quiet for a moment. Then she said, "But that word do hurt other peoples."

CHAPTER 29

Carville was strange in many ways, not the least of which was the use of aliases. No one, it seemed, used his or her real name.

Inmates were issued numbers by the guards, as well as nicknames by Link and his friends. Link had called me Clark so often that most of the leprosy patients and inmates actually thought it was my real name. That was fine with me—a nickname that was the secret identity of a superhero who posed as a journalist. Father Reynolds had taken a new name when he entered the Franciscan monastery. All the Sisters of Charity were given new Christian names when they made their vows to chastity and poverty.

And the leprosy patients took on pseudonyms when they arrived. Jimmy Harris had used J. T. Holcomb for decades; Ann Page picked her name from the label of a jelly jar, and a Texas beauty queen chose Molly for her first name, after her prizewinning cow.

Though most of the inmates thought Ella's name was Cella, she never did take an alias.

"Didn't see no need to change," she said. "Ella the name my momma gave me." Then she asked, "You want to change your name?"

No, I told her, but I wouldn't mind a new Social Security number.

Monikers at Carville were not restricted to individuals' names. Leprosy had all sorts of aliases, too. Most patients almost never spoke the word *leprosy*. Among themselves, they referred to it as "the disease," "the package," or "the gazeek." In fact, the patients had launched an international campaign to rename leprosy.

In 1931, Sydney Levyson, a pharmacist from Texas, was shipped to Carville by railcar. Sydney was a handsome, stylish young man. With his topcoat buttoned and leather bag in hand, he sat inside the locked car as the couriers talked about the "leper" inside.

Sydney, like most arrivals, took a new name. He would forever be known as Stanley Stein. Though he had been diagnosed with a dreaded affliction, being called a "leper"—and all that came with the word—didn't sit well with Stanley. He decided to do something about it.

He launched a publication entitled *The Star*. Its purpose: to eradicate the use of the words *leprosy* and *leper*. The stigma associated with leprosy was so ingrained in society, Stein argued, that nothing short of a name change could lift it. Stein and crew at *The Star* promoted a new label named after Armauer Hansen, the Norwegian scientist who discovered the bacteria that caused leprosy. The slogan on the cover of each edition of *The Star* read: Radiating the Light of Truth on Hansen's Disease. They made great progress in their campaign among health-care workers, victims of the disease, and their advocates, but they seriously underestimated public paranoia. And ignorance.

In many ways, the renaming actually fostered fear and stigma. When uninformed citizens discovered that a Hansen's disease patient was afflicted with a disorder formerly known as leprosy, they felt duped. The stigma was transferred to the new term. Their paranoia and suspicions were fueled by what they perceived to be a covert act—hiding the true identity of the ancient disease.

CHAPTER 30

Doc had one close friend at Carville, Dan Duchaine. Dan was in his late thirties. His short spiky hair was beginning to gray at the temples. He also had no body hair. He shaved his entire body every couple of days. Dan was a nationally recognized steroid and bodybuilding expert who knew as much about physiology as Doc. Duchaine and Doc would sit for hours and discuss mitochondrial biogenesis and weight lifting, varying reactions in type I and type II muscle fiber, genetics and human endurance, caloric consumption and heat. Dan, who for most of his career had promoted steroids and supplements, had an interest in Doc's heat pill, especially its potential for burning fat off bodybuilders.

Dan was a superstar in the bodybuilding industry. He started out as just another bodybuilder, but he didn't have the genetic makeup to compete on a national level. What his body lacked, he made up for with brains. In 1988, Dan instructed major Olympic athletes in techniques for passing drug testing. He was so adept, authorities designed tests and regulations and programs with Dan in mind. When he introduced a new supplement to the marketplace, the FDA wasn't far behind to prohibit its use. Dan had written four books, including *The Underground Steroid Handbook*, which sold thousands of copies among bodybuilders.

Dan was in the midst of another project, creating an audiotape of interviews with "The Steroid Guru" on diet, supplements, weight lifting, and exercise. A friend recorded the interviews while Dan answered questions on the pay telephones inside the prison.

One afternoon in the inmate courtyard, I asked Dan how they dealt with the recorded message that interrupted telephone conversations from Carville.

"We leave them in," Dan said. "We can charge more."

Dan understood the value of inside information. The bodybuilding world—at least the bodybuilders who used steroids—knew their guru was imprisoned. But he couldn't be silenced. Much like Doc, who spent hours each day reading the latest developments in medicine, Dan put his prison time to good use. He had dozens of inmates who were more than happy to serve as his guinea pigs. They ate exactly what Dan told them to eat. They lifted weights using his techniques. They followed his instructions on when and how to exercise. Then, for a fee, Dan reported the results of his studies over the prison phone lines. From a marketing perspective, it was brilliant.

Dan had suffered a stroke a few years ago, when he was in his mid-thirties. That's why he'd been moved to Carville. Except for stiffness in his left arm and a slight hesitation in his speech, the stroke hadn't done much damage.

Dan scoffed at the idea that steroids might have contributed to his condition. To me, he sounded defensive, but I wasn't the expert.

On a Saturday afternoon when Dan and Doc took a break from sharing theories on the most efficient methods to enhance energy consumption at a cellular level, I asked them if they knew about the attempts to rename leprosy.

"Yeah, Hansen's disease," Doc said.

"It didn't catch on," I said. "What should we call them?"

"Lepers!" Dan yelled.

I ignored the remark and pointed out that most of the patients didn't have an active bacterial infection. So calling them "patients" didn't seem right either.

Duchaine suggested, for accuracy and political correctness, that we call them *carbon-based units with residual effects of ancient bacterial infection.*

"It's really a public relations issue," I said. I told Doc and Dan that I wanted to help Ella and Harry and Annie Ruth. "They're good people. They don't deserve this."

Doc rolled his eyes. He told Dan that I had fallen prey to their *heart-wrenching* stories. Doc looked over at Duchaine. "He believes them."

Duchaine was never short on opinions or advice. He was happy to share his thoughts on just about any topic. "They lost their childhoods," he said, pausing for a moment to reflect. "It's not uncommon for them to create new histories, even new identities."

Doc raised an eyebrow. "See?"

I left the two of them to their science and cynicism and walked the track. I thought about whether Ella and Jimmy and Annie Ruth would really create histories for themselves. I had no way of knowing. But the thought occurred to me that Doc and Dan had been in jail for years. Maybe I should be questioning their stories.

I walked until the sun started to set. Then I went back to my room, propped a pillow against the wall at the end of my bed, and picked up a book.

Link, who had just finished an all-day game of spades, walked into the room.

"What that book?" he asked.

"A book my mother sent me."

"What it called?"

I closed the book and read the title aloud: "*Pleasing You Is Destroying Me.*"

Link threw his head back. "Man, you is so white!" Link had a point. "Why you white people read books to solve your problems?"

"You might be surprised," I said. "Books can change your life."

"Where was that book when you was robbing them banks!?"

Doc, who rarely spoke to Link, added, "Neil wants to save the world. And help the patients."

I couldn't believe Doc was siding with Link. I looked at both of them, trying not to sound too earnest, and said, "It feels good to help people."

"If I need some money," Link said, "it'll make you feel good to give me some? 'Cause I need seven dollars."

My father had just put $100 in my inmate account. I walked over to my locker and counted out twenty-eight quarters.

"I'll give you this, if you promise one thing." Link waited for the caveat. "Promise," I said, "that you will never carjack when there's a kid in the car."

"All right," Link said, grabbing my quarters, "and I'll pay you back if you promise you won't rob no more banks."

A pebble hit our window. I looked out. It was Smeltzer. I pushed the window up and looked out. Smeltzer reached into his coat pocket and pulled out two Ziploc bags, one full of chicken wings, the other pork chops. Smeltzer stretched out his arms and dangled the bags. "Hungry?" he asked. Food smuggled by a leprosy patient didn't do much for my appetite. "I have a newspaper, too," he said. I said no thanks, and he told me to get my roommates. Link took my place in the window.

"How much?" Link asked. Smeltzer said he could get both packages for two dollars. Link counted out eight quarters from the money I had just loaned him.

"You lepers is makin' too much money off us," Link said. I looked at Link and shook my head. He smiled, like he was proud to have used the term correctly.

"Don't call me that!" Smeltzer yelled.

"You got leprosy, don't you?" Link said. "What the fuck you want us to call you?"

"I don't have the same disease that was in the Bible," Smeltzer yelled. "Don't ever call me that or I'll report you to the guards."

I suddenly realized we should be reporting Smeltzer. He wasn't supposed to be on the inmate side in the first place.

Link asked him again, "So what the fuck you want us to call you?"

Smeltzer hesitated. Even the patients couldn't agree among themselves on an appropriate label. Some used "Hansen's disease patient," a mouthful. Others simply wanted to be called "residents." "I don't know," he said after a while, "maybe Hansenite."

Sounded like luggage to me.

Link leaned out of the window and dropped the quarters. Smeltzer tossed the Ziploc bags to Link, who struggled back into the room, food in hand. Link put a paper towel on my bed and spread out the chicken wings.

Doc yelled out to Smeltzer, "Next time, bring some lemons!"

CHAPTER 31

As the leaves started to turn on the trees, Linda and the kids traveled to Oxford to stay with her family. I missed our family visit, but I was happy she and the kids could be away from the prison visiting room for a while.

Doc thought my eagerness to help the leprosy patients was ludicrous, but I was growing closer to them. They were becoming a family away from home. Especially Ella.

I walked the track and imagined what I would do if I were in charge of altering public opinion, if I were editor of *The Star*. The patients needed some good, old-fashioned public relations work. "Hansen's disease" was never going to catch on. And the problem with the label "leprosy" was that people were blinded by their own preconceived notions before they had a chance to learn anything factual.

In contrast, Lou Gehrig's disease was named for a great ballplayer, a guy whom everyone admired. Maybe that was what leprosy needed. A more positive twist.

As I walked in circles around the track, I started to brainstorm. Something like *Crusaders' disease* or *Lazarus syndrome* sounded noble enough, but it might also bring up frightful images such as scenes from *Ben-Hur*. *Damien's disorder* would have been a perfect name before *The Omen* movies came out, but now people would think about the Antichrist and that could adversely add to the stigma. Then I contemplated the possibility of names that acknowledged leprosy's ancient roots, and its place as the oldest malady known to man. *The Holy disease* looked fine on paper but phonetically could be misconstrued. *Assisi complex* was not a bad possibility unless you had a lisp.

Clearly, none of these was the answer, but solutions never came early in the creative process. I would continue to experiment with possibilities while I gathered stories for my exposé.

One afternoon while Steve Read and I walked to Father Reynolds's study, I ran into Janet Cedars, a woman with a remarkable story. In the 1950s, after she had just been elected head cheerleader at her high school in Texas, she was diagnosed with leprosy and sent to Carville. While here, she fell in love, married, and had children. But her children couldn't stay with her. They were placed with a family member in Texas. A staff member told me that for the child's christening, the foster parents brought the baby back to Carville so Janet could see the ceremony. But Janet wasn't allowed to touch the child. Janet was one of the last to suffer from this kind of treatment. In 1962, free of the disease, she was reunited with her family. She completed college and lived a normal life outside of Carville. Then, in an extraordinary move, Janet was hired as a teacher, and later as the director of public affairs at Carville. She was held up as an example of how much things had changed. The Public Health Service made an effort to publicly promote Janet's transition from patient to employee. The inmates had all heard her story, and her name appeared in the newspapers pretty regularly.

No one would ever have guessed she had suffered from the disease. Her face was flawless, as beautiful now as it had been when she was a cheerleader, though a bit more round. I noticed a file folder clutched in her arms. It was labeled Centennial Celebration.

"Mrs. Cedars," I said, as I approached her.

"Yes," she said, a bit surprised I knew her name. "May I help you?"

"I've been thinking about some of the public relations issues you face. I have some ideas I'd love to talk with you about."

She looked at me, standing in my prison uniform, not able to hold on to my own freedom. "I did this sort of thing on the outside. Are you familiar with *Louisiana Life* magazine?" I explained that my concepts for rebranding leprosy weren't fully formulated, but I was sure they were headed in the right direction.

She stared at me while I held forth on how best to alter public opinion. Then she nodded and repositioned her file folder.

As she walked away, I called out, "I look forward to the centennial celebration. And let me know if I can help."

Janet glanced back over her shoulder and scurried away.

"You moron!" Steve said. "I bet she'll run right out and call a press conference. *Oh, yes, he's a very insightful federal convict,*" Steve said, channeling Janet, as if she were pitching my services to a group. "*His company? Oh, I think it's called 'PR with Conviction.' He doesn't have a telephone, but he does have an inmate number. I think we should just turn the entire program over to him.*" Then Steve went back to his own sarcastic voice. "Oh! I've got your slogan: 'Captive audiences are our specialty!'"

An inmate offering professional services might seem a bit odd, but Janet had been imprisoned, too. I thought she would understand. And I thought I had something to offer.

CHAPTER 32

Smeltzer's efforts to profit from the inmates reached a fever pitch in October. His newest scheme involved an inmate muffuletta feast. Muffulettas—enormous, wheel-sized Italian sandwiches stuffed with meats, cheeses, and sauces and topped with an olive salad— were a New Orleans specialty. And the best place to get muffulettas in New Orleans was Central Grocery, an old-time store located in the French Quarter.

Smeltzer had collected $4 from forty different inmates. He promised the prisoners a bounty like no captive men had ever seen. On the afternoon of the clandestine event, Smeltzer's cousin, a New Orleans native, purchased ten muffulettas from Central Grocery. The sandwiches were cut into quarters and then wrapped in heavy-duty aluminum foil. Smeltzer's cousin packed the muffulettas in a blanket and then stuffed them inside a black garbage bag. He drove from New Orleans to the edge of the colony where he dropped the bag at the hole in the fence. Smeltzer was waiting. He carried the bag of sandwiches to the rubbish collection station, where Dog Man, the inmate fond of howling, waited in his motorized garbage cart. Smeltzer placed the plastic bag in the back of the cart along with the other trash. Dog Man drove the cart, like he did every afternoon, toward the incinerator. But on this day, he stopped at a window at St. Amant, the inmate dorm closest to the leprosy side. Dog Man signaled the inmates inside with a series of predetermined yelps and barks. A window was opened. Dog Man pushed the bag through, and forty inmates feasted on still-warm muffulettas.

I didn't join in the bounty, but I loaned Link $4 so he could buy in. Frankly, I was afraid to be a part of the operation. If inmates could successfully smuggle a garbage bag full of food, albeit with the help of leprosy patients, I wondered what else could have made it through the gates. Drugs, tattoo guns, knives.

Every inmate involved had pledged not to speak of the operation, but I knew all too well that inmates had a hard time keeping secrets, especially Link. I also knew if the guards found out, everyone involved would get additional time tacked onto their sentence. No amount of food or fun was worth keeping me from getting home as soon as possible.

"Goddamn," Link said, the morning after the feast. "That motherfucker made $160 off them muffulettas. What a leper gonna do with all that money."

"Seems reasonable," I said, reminding Link that the muffulettas were delivered hot from New Orleans. "We just have a skewed perception of money right now."

"Screwed what!?"

I tried explaining that the $20 limit—two rolls of quarters—imposed on inmates made things *appear* more valuable, but Link lost interest in my explanation.

I had to admit, my perception of money had shifted, too. Now, I found it hard to believe that two years earlier, in the summer of 1991, I had finalized the $1.2 million purchase of *Louisiana Life* (the first step in my five-year plan) and moved my corporate offices into a $4,000-per-month penthouse suite in the old Markham Hotel building. My office was once a grand ballroom where Gulfport society danced on marble and men leaned against mahogany walls smoking cigars. It was the very place my grandmother and grandfather met at a spring ball in 1938.

At that time, she was among the wealthiest single women in the United States. Her grandfather, an early industrialist who had financial interests in anything made of iron or steel, was one of the men who financed young Henry Ford's automotive venture. Later, he built a company to supply parts to the burgeoning auto industry. It became the Borg-Warner Corporation.

My grandmother's father died in the 1918 flu pandemic when she was a baby. She had no memory of him. The only reminder was the fortune she inherited.

In 1938, my grandmother, Martha Johnson, was a short, muscular, athletic nineteen-year-old college girl and she was painfully shy. Despite all she could buy, beauty was unattainable for her. But she was generous and kind.

On a Friday evening that May, on a break from Vassar, she went to the spring dance at the top of the glamorous Markham Hotel building. She sat alone in a chair next to the mahogany walls until a handsome young man, on a dare from a friend, asked her to dance. They spent the evening together dancing on the marble floor and drinking from the boy's silver flask. Martha was enamored with his charm and wit. In the early morning hours, in a fog of alcohol and euphoria, Martha and the boy drove across the state line and married.

The next morning, this young man—my grandfather—not yet nineteen, awoke hungover, next to Martha. And suddenly, nothing was out of reach.

They threw lavish parties and bought a mansion with a huge swimming pool. They spent their summers at a home on Walloon Lake, Michigan, and never missed a party at the country club. Their live-in servants, Wash and Esther, traveled wherever they ventured to care for their growing family.

To an outsider, my grandparents' life must have been something to envy. Money was not an issue, and my grandfather did not need to work. Instead, he went on a thirty-year odyssey in search of adventure. It started with excessive parties, reckless investments, and extravagant purchases, like his fifty-foot yacht, *The Weekender*. As the years passed, adventure became more elusive. He turned to alcohol, gambling, and women.

The family money had survived the Depression, but the wealth couldn't bear the strain of my grandfather. When the money ran out in the 1960s, he divorced Martha and married a woman almost twenty years younger.

Martha's life unraveled. She spent her remaining funds on a reconditioned riverboat that drove her to the verge of bankruptcy. She was committed to the state mental hospital in Whitfield. During her holiday stay with us, she told me how much she liked the people at the institution.

"They're nice to me," she said.

In the last decade of my grandmother's life, she waited in line at free medical clinics; she received a small welfare check; she worked part-time at Goodwill Industries as a clerk; and she rented a government-subsidized apartment that smelled like cats. Old photo albums lined her cheap bookcase. Reminders of a privileged time.

Fifty-three years after the spring dance of 1938, I worked each day in an office not more than a few feet away from where my grandfather first saw my grandmother. I liked the symmetry. As I planned my rise in the world of publishing, I vowed to bring our family back to its rightful spot. And I was determined to do it fast.

CHAPTER 33

Because the leprosy patients liked my menu board illustrations, the guards gave me another job: garnish man. They also gave me a *Better Homes and Gardens* garnish book that illustrated the most up-to-date garnishes being used by the finest caterers in America. I took the book back to my room and read it at night in preparation for my new duties.

My first day on the job, I ran into a problem. I had spent hours studying the garnish book, and I felt ready to prepare a centerpiece like Carville had never seen. But I was missing the most important tool. A knife. Inmates weren't allowed to have knives. I tried carving into some fruit with a plastic knife from the cafeteria, but it was useless.

I asked the guard on duty how I was supposed to prepare garnish without cutlery.

"You can check out a knife," he said. He led me to a closet locked with a dead bolt. Inside, there was a small cage welded to the wall. The cage was secured with a padlock. Inside, two large magnetic strips held a dozen sharp knives with black plastic handles. Each knife was marked with a number, a slapdash, hand-drawn digit that looked like it had been written with Liquid Paper.

"I could rewrite those numbers, if you like," I told the guard. "I could make them neater." He ignored my offer, opened the cage, and handed me knife number 6, a small paring knife. He wrote my name and inmate number on a clipboard, said the knife was my responsibility, and added that if it turned up missing, I'd be put in the hole for thirty days. I made a mental note not to leave the knife lying around the kitchen.

My garnish station was just outside the produce cooler on the leprosy patient side. Two large stainless steel tables were dedicated for garnish. I was the only inmate allowed to enter, though I could hear the conversations of Chase and Lonnie, the two inmates in charge of the food warehouse. They had been at Carville since the day the prison opened. If the federal prisons had had a trusty system in place, Chase and Lonnie would have held the distinction of being trusties. The two of them ordered the food for the cafeteria. They drove trucks. They held shrimp boils on holidays. They had the run of the place. As I carved into the fruit and built the foundations for my first set of garnishes, I listened to Chase and Lonnie discuss flaws in the plotline of *Gilligan's Island*, explore the dangers of spitting on a guard's food, and engage in a debate about whether a horse will fall in love with a human who urinates on its feed.

The garnish business was messy. I had always hated to get my hands dirty. As a husband charged with changing a flat tire or hollowing a pumpkin or potting a plant, I would get agitated. I performed the tasks hurriedly, with trepidation, tight-lipped, nostrils flared. I worked fast so I could get to soap and water quickly. But it was more than the dirt that bothered me. I liked the way my hands smelled after I had applied cologne. Each morning, I dabbed the backs of my hands so they would smell nice whenever my hands passed anywhere near my nose, or that of another. Washing my hands to rid them of dirt also rid them of the scent I so loved.

I wasn't able to cut into the fruit without getting my hands covered in sugary, sticky juice. If I stopped to wash after every cut, I would never finish. So I decided to be outrageous. I lost myself in the carving of cantaloupes, strawberries, peaches, and watermelon. I felt like a child again, making mud pies or building a sandcastle. The garnish book's step-by-step instructions were brilliant. And simple. On my first attempt, I made a rather elegant swan from a honeydew melon. Its long, curved neck looked as if it could reach around to clean the tall feathers that stretched upward from its wings. As a companion for the swan, I carved a duck from lemons. The duck's webbed feet were made from intricately notched carrot slices. Then I made a series of islands. For sand, I halved a baking potato. I made shallow cuts into a carrot stick, transforming it into the trunk of a palm tree. I added long, curved slivers of green peppers to make perfect palm branches. Then, from an oddly shaped watermelon, I pieced together the head of an alligator with an open mouth and sharp teeth. I thought it would make a great centerpiece to adorn the salad bar.

I was starting to like the garnish business. I understood why the guards would prepare a garnish for the leprosy patient cafeteria, but it seemed odd they would bother to decorate a salad bar for prisoners. Then again, it seemed pretty odd that prisoners had a salad bar.

As I became more familiar with the traits of certain fruits and vegetables, I experimented with combinations of color and texture. Purple-headed lettuce, cut with a certain precision, opened like an exotic flower. When accented with maraschino cherries, it took on the traits of a rare, insect-eating blossom. Soon, using cantaloupes and oddly shaped gourds, I learned to sculpt the likenesses of certain inmates, and even guards, complete with bad hair, bulging eyes, and badges.

Ella and the other patients said the cafeteria line looked like a formal buffet. It gave them something new to look forward to each day. My creations got lots of compliments on the patient side. Harry stopped to take a look and stuck his thumb, his only fully intact digit, in the air to show his approval.

As I carried the last of the trays out of the kitchen, an inmate stopped to admire my work.

"That's gorgeous," he said in a Cajun accent. "Did you do that on the outside?"

"Yes," I told him, "I'm in here for garnishing with reckless abandon." In a way, it was true. I was good at the job. Polish, shine, attention to detail, and the appearance of perfection were my forte. It came quite naturally. I'd had plenty of practice on the outside.

My financial statement for *Coast Magazine* was a thing of beauty. I used advanced publishing software to design it. I selected an old-fashioned font—Baskerville—and adjusted the tracking so the numbers aligned in a way no accounting software could match. My software didn't actually add numbers, but it enabled me to produce a specimen the likes of which the banks had never seen. I printed the financials on custom-made, cotton-fiber stationery for the bankers and investors who requested the statement. It looked and smelled and felt like affluence, almost too good to be true.

Before acquiring *Louisiana Life*, kiting checks was an occasional financing technique. But now, I watched the clock carefully. At 2:00 P.M. each day the banks would collect the last deposits. On any given day, I would check the mail for advertising payments, chat with investors about their next installment, and explore sources of alternative financing. Then I would calculate any cash shortfall and prepare a covert transfer.

December 21, 1991, was a typical day. I sat at my desk and wrote two checks. One for $89,000; the other for $118,500. The latter probably didn't need to be quite so high, but I thought it better to be safe than to fall into the insufficient funds realm and the attention that would bring. Both checks were written from my company; both were made out to my company. But the checks were drawn on different banks.

I wrote a tiny note at the bottom of my daily planner. A reminder of what I would need to cover tomorrow. I put off, for a moment, the preparation of the deposit slips. I wanted to arrive at the banks as close to 2:00 P.M. as possible. I wanted the bookkeeping people to be in a hurry, to be so absorbed, so intent on meeting their cutoff time that they wouldn't question the size of the checks.

I passed a few moments standing at the corner window of my ninth-floor office. The view of the Mississippi coastline was unobstructed. South, I could see the barrier islands more than ten miles out into the Gulf of Mexico. East, a stretch of man-made beach, bordered by live oaks and pines, curved along the Gulf of Mexico coastline. The walls of my office were lined with covers from vintage *Life* magazines, a tribute to one of my idols, Henry Luce, and a reminder of the power that goes along with media ownership.

My leather-topped desk, a gift from my father, was covered with a carpet of work—page proofs from the new issue, a stack of letters, my planner, and two large corporate checkbooks.

The stack of letters on my desk included our nomination as Small Business of the Year, a request to speak to a civic club, a few notes from parishioners at St. Peter's by-the-Sea Episcopal Church where I served as senior warden, an invitation to serve on the board of directors of the Coast Anti-Crime Commission, and a letter from the school board president who had agreed to let me send a young undercover reporter to pose as a student at Gulfport High School.

I didn't have time to respond to the correspondence. The antique clock on my wall read 1:40 P.M. Twenty minutes until the banks would post.

I prepared the two deposits, one for each bank. I mingled the large checks among smaller subscription payments, as if they were just another in a line of routine deposits. I waved to employees as I passed by their offices and cubicles, making sure to smile an encouraging smile to my salespeople, and called out to no one in particular that I was off to the bank.

At the elevator, I noticed the recently polished copper-and-glass mail chute, a still-operational holdover from the building's hotel days, and the perfectly reflective brass doors of the elevator. I placed the two deposit books under my arm and straightened my tie. Every afternoon, a few minutes before two o'clock, I waited for the elevator to make its way to the ninth floor and examined my reflection. My hair was beginning to turn gray, but I really didn't mind. I actually wanted more. It added, I thought, an air of stability and soundness. Maybe even prudence.

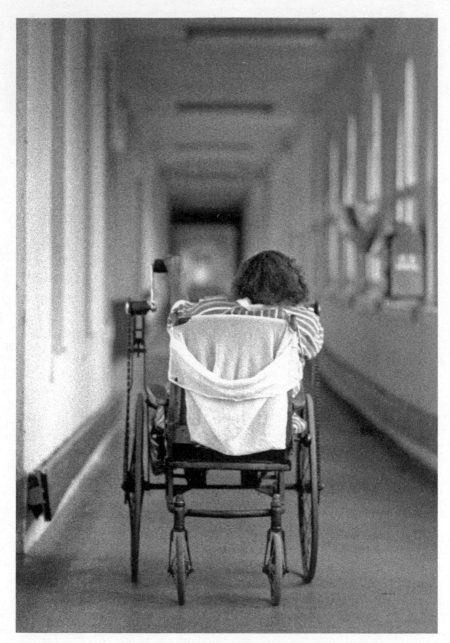

Ella Bounds meanders down the corridor.

CHAPTER 34

The prison was quiet and cool the day I turned thirty-three. My mother sent me several books, and my father put $100 in my commissary account. Neil and Maggie sent homemade birthday cards.

On the same day, Doc also got some good news. He was awarded a U.S. patent for his medical injection device designed to cure impotence. The one-page document contained an illustration of the injector with a curved base for a snug fit around the penis. But, to me, that wasn't the most interesting part. The document listed the "inventor" as Victor Dombrowsky. It even listed his federal inmate number.

Doc was a brilliant physician and innovator. But he didn't pursue medical ventures unless there was potential for great profit. When it came to money, Doc went for the jugular. He dispensed heat pills to the obese and invented a cure for impotent men. No two groups would be more willing to pay any price to be cured.

Thirty years earlier, another trailblazing physician lived at Carville, Dr. Paul Brand. Dr. Brand was the antithesis to Doc.

Brand, the son of missionaries, grew up devoutly religious in India. After completing his medical training in Britain, Dr. Brand returned to India to follow in his parents' footsteps. When he first encountered leprosy patients, he was reluctant to treat them for fear of contagion. But then he realized that few, if any, qualified physicians worked with the "lepers." Their care was provided by religious zealots or monks.

With the help of his faith, Paul Brand overcame his own fears. During the time he worked in India, his knowledge of, and skills in, the orthopedic aspects of the disease surpassed any physician's before him. He developed groundbreaking techniques for restoring nerve-damaged hands. But it wasn't always appreciated by the leprosy patients. His successful surgeries created a financial hardship. Begging was their only source of income. With corrected, functional hands, their ability to collect alms was greatly diminished.

After two decades of selfless service in India, Dr. Brand accepted a position at Carville.

The patients called him "Saint Paul." Brand treated each patient with the utmost respect. He wanted to help patients regain their dignity, as well as repair their bodies. One Carville patient said that Dr. Brand touched her foot as if he were "handling a delicate, but broken instrument."

Better than anyone else, Brand understood the curse of insensitive limbs. Sores on numb feet went undetected. Injuries to anesthetized hands could go unnoticed for weeks. He had perfected techniques and surgeries to repair his patients' damaged hands and feet.

In the years before his arrival, amputations at Carville were commonplace. But in the first seven years he served as lead orthopedic surgeon, not a single amputation took place.

Ella rolled into the cafeteria wearing her prosthetic legs.

"You wear those because it's my birthday?" I asked.

Ella knocked on one of her artificial legs and smiled. "How old you is?"

"Thirty-three," I told her.

"Jesus that old when he rise up," she said.

All good southerners knew Jesus had died at age thirty-three. I figured Ella would recount the lessons she had heard in church about the thirty-third year being one of change and transformation for Christians, but she didn't.

So I told her a story my mother recounted on birthdays. It was my favorite family story. A great tale of our Scottish heritage.

The MacNeil clan chieftain, the story went, was feuding with another clan leader over ownership of an island. Rather than risk widespread death and injury, the two agreed to settle the issue in a competition. The two chiefs would race from the mainland to the island, and the man who first put his hand upon land would claim the isle for his clan. In the early morning fog, the two men set out in their open boats. The race was a dead heat until the very end, when the opposing clan chief inched ahead. At the end of the race, with MacNeil a few boat lengths behind, the island looked lost for the MacNeils. As the other chief stepped out of his boat in the shallow water near shore, the MacNeil chieftain removed his blade and severed his right hand. He picked up the bloody stub and hurled it onto shore, winning the island for the MacNeils.

I loved hearing the story as much as my mother loved telling it. Afterward, she would point to the red hand on the coat of arms that decorated our wall and recite the Latin motto printed under the crest—*Vincere Vel Mori*. "To Conquer or Die."

"That 'splains a thing or two," Ella said.

Just as I finished telling the story, I looked at Ella's legs, and I realized she might not want to hear a story about a man who performs his own amputation. I tried to change the subject.

"Do you remember your thirty-third birthday?"

"Sure do," she said. "Got these." Ella tapped on her prosthetic legs again. "Looks like they was made for a white woman." Ella laughed, but she was right. The light color didn't come close to matching her skin.

Ella didn't want to miss anything. When she was young, she loved to run and dance. She stayed on the move. And over the years, her legs were badly damaged, but Ella didn't feel the pain. In the late 1940s, when she turned thirty-three, her legs were taken away just above the knee. Paul Brand arrived at Carville twenty years too late for Ella.

Her surgery forever altered the way she looked and the way she lived. Her body was transformed. And for nearly fifty years, she had meandered the halls in her hand-cranked wheelchair—welcoming new patients, visitors, even inmates; taking drinks to the residents who couldn't make it to the canteen on their own; smiling at everyone. I couldn't imagine her any other way.

At thirty-three, Ella started over.

Stan and Sarah, the blind couple who tapped their way through the corridors.

CHAPTER 35

During the five months I'd been at Carville, I had talked to almost all the patients who ate breakfast in the cafeteria. I'd made some good friends. However, I had never approached Stan and Sarah, the blind couple. I couldn't convey with a smile or a nod that I wasn't a dangerous criminal. They couldn't see me, but I had watched them as they walked arm in arm around the colony. Most blind people tapped their cane lightly, but Stan slapped his stick against the floor and walls to send the vibration past his numbed hand to his arm and shoulder. I could hear them coming down the corridors before I ever saw them. They maneuvered fairly well in the cafeteria, at church, and in the hallways, but on occasion, Stan became disoriented.

After lunch one afternoon, when I was alone with them in the patient cafeteria, Stan, with Sarah on his arm, tapped his way around the tables and chairs. They had almost made it to the door leading to the leprosy patient dormitories when he veered off course. Stan ushered Sarah into an empty nook. A coat closet. I watched for a moment. I noticed the confusion on his face. He turned and guided his wife into a wall. Then, he turned, and tapped his way back into the corner again. I couldn't stand it. I walked over and softly gripped his tapping arm.

"This way," I said, gently tugging Stan's elbow, leading him toward the exit.

"Don't touch me!" he snapped. "We don't want your help."

Stan jerked his arm out of my grasp. I stepped back into the hallway. Stan gathered his bearings and led his wife out of the alcove.

They tapped their way to the exit and turned left toward their dormitory. I stood alone in the cafeteria listening to the sound of his stick against the concrete. It faded away, and I sat down.

Stan and Sarah couldn't see me. They didn't see my expression. They didn't know I was trying to help. All they knew was that I was a convict. They were afraid of me. And for the first time, I understood how the leprosy patients must have felt when people were afraid of their touch.

Me in Oxford, Mississippi, just after the fall that left scars on my forehead, 1961.

CHAPTER 36

As I immersed myself in reporting on the patients, my reaction to their deformities changed in ways I never could have imagined. The shortened fingers of a patient from Trinidad were perfectly smooth and symmetrical. At times when I saw him talk and gesture with his miniature hands, he looked like a magical being who didn't have to bother with human traits like fingernails that needed to be cleaned or clipped or groomed. His hands were nothing short of perfect. For him.

I had grown accustomed to Harry's distorted voice. When he would reach deep into his front pocket to retrieve his wallet and say, "My mudder taut me to do dat," I heard him clearly. His specially designed Velcro shoes fit his unique feet in a way that made standard shoes seem like restrictive boxes. His tools—from a device to help him button his shirt to the utensils he used to eat—didn't seem unusual anymore. And his incomparable hands. The white skin under his palm met the dark skin from the back of his hand to form a seam where his three middle fingers once existed. His hands were one of a kind. White circles covered the knuckles where his left thumb and index finger once were, as if the pigmentation had been rubbed off. I imagined no other man or woman on earth had hands quite like his. The more I saw them, the more comfortable I became.

Ella was no different. I couldn't imagine her without her wheelchair. The only time she seemed odd was when she put on her prosthetic legs. I was so accustomed to the way her dress fell across the front of her chair, the way her hands gripped the handles of her cranks, and the way her wheelchair wobbled as if it were the seasoned gait of any other nondisabled woman in her eighties. Her deformities disappeared.

I had experienced this before—at the other end of the spectrum. When I dated the homecoming queen at Ole Miss, I was at first astonished by her loveliness. To see her walk across the Ole Miss campus, a beauty among some of the most beautiful women in the world, would make me light-headed. At times, I couldn't believe she was attracted to me. I watched people stop to stare. Men couldn't help turning their heads. Sometimes women did, too. She possessed the kind of physical perfection that seemed almost unfair. But as our relationship progressed, my awe of her waned. Her appearance had not changed, but I ceased to be dizzy when I saw her. Her perfect nose and lips and hair and eyes were as mundane as the features on my own face.

Intimate, prolonged contact, it seemed, made everything commonplace. Beauty *and* disfigurement disappeared with familiarity. Beauty queens became ordinary; leprosy patients did, too.

I had spent my life surrounding myself with beautiful people. And I made certain no one ever recognized my shortcomings. A childhood accident left two oblong scars in the center of my forehead. Just after my first birthday, while my babysitter talked on the phone, I tumbled down five concrete steps at Avent Acres apartments in Oxford (where my family was living while my father attended law school at Ole Miss). The injuries weren't the kind of cuts that stitches could repair; the rough edge of the concrete scraped the skin off my forehead. My mother called it concrete burn. A physician covered the injuries with gauze and sent us home, but a week later, he pulled off the gauze and with it came the two scabs. Two oblong scars now dominated my forehead.

Until I was almost ten years old, I was oblivious to the peculiar marks on my forehead. But when I finally did recognize their oddity, I went to extraordinary lengths to hide them. I kept my bangs long enough to cover my forehead, and I constantly pressed my hair down to ensure the scars were hidden. I lowered my head on windy days so no one would get a glimpse if a strong gust came along. During swim team practice, I was careful to emerge from underwater perfectly perpendicular so my hair would never be swept back to reveal my flaw. During school pictures, photographers would come at me with a comb or brush, but I would press my hands over my bangs and tell them I liked my hair this way.

At home, locked inside my bathroom, I would push my hair back and stare at the unsightly scars. Keeping my forehead constantly covered was inconvenient, but the alternative was unthinkable. Others might find the blemishes abnormal. Maybe even repulsive. In front of the mirror, I resolved I would never let anyone see my damaged skin.

In the summer of 1980, after I made good grades my freshman year at Ole Miss, my father arranged to have the scars repaired by a plastic surgeon. During the preliminary exam, the doctor pinched and prodded the scar tissue and assured me that when he was finished the hairline scars left behind would blend in perfectly with the wrinkles that formed in my forehead when I raised my eyebrows. The procedure took less than thirty minutes. I was awake the entire time. When he finished cutting out the oval scar tissue and pulled the skin together with the final sutures, he gently pressed a piece of gauze over the incisions.

"That isn't going to stick to the scab, is it?"

He held up a tube of ointment and instructed me to apply it to the scars twice a day.

For the first time in years, I felt free of the humiliation of being different, and I felt light because I had left my ugliness behind.

But this imprisonment and the label "ex-con" would follow me for life.

I had no idea how people would treat me, or if they would embrace me, living with my flaws exposed. I'd never given anyone the chance.

I didn't tell Ella that her physical defects were disappearing for me, but I did tell her about how much time and energy I had put into maintaining my image, how much money I had spent making an impression, how many people I had hurt, and how much stress had come with creating an illusion.

"I wanted everyone to think I was perfect," I said, as if I'd had some great insight.

"Well," Ella said, "you ain't gotta worry about that no more."

She was right. My conviction had been splashed across the front pages of newspapers. It had been the top story of the evening news. The scandal rocked my hometown. The behaviors I had developed to hide business setbacks, cash shortfalls, and any hint of failure were exposed on April 9, 1992. A Thursday.

The *Coast Magazine* offices were unusually quiet. And empty. All twenty-nine employees were attending an advertising sales symposium. Helen Berman, a sales expert from Washington state, had flown in to lead a motivational seminar for my staff.

I received a phone call from Albert Dane, my loan officer at Hancock Bank. Albert had helped our company grow from the very beginning. He had arranged for equipment leases, short-term loans, and credit card processing. He was a friend and supporter of our business.

"You need to come down to the bank," Albert said. I looked at the clock. The bank had just opened for the morning. I asked if there were some problem. "You just need to come down here," he repeated.

I packed my briefcase and checked the notes I'd made the day before at the bottom of my planner—a $419,000 check written on the Hancock Bank account; $345,000 drawn on Peoples Bank. I hoped some glitch had come to the bank's attention that could easily be resolved. But I knew better. Albert wouldn't call if the news weren't awful.

I looked out of my window at the panorama of the coast. I stood for a moment. The beach and gulf were bright and clear in the morning light.

On my walk to Hancock Bank, I passed the offices where my father and his grandfather had practiced law. Two blocks down was where I had visited my grandfather in the purchasing office of Mississippi Power. Five generations of my family had lived and worked in Gulfport as attorneys, teachers, surveyors, and entrepreneurs. This was my town, and I had hoped to be its greatest champion. My magazine depicted the region without its blemishes or flaws, like a magic mirror that reflected away any imperfections.

I walked slowly. I wanted to remember how Gulfport felt, and how the town felt about me. I had made many friends. I had made them look good in my magazines. Of course, it had ultimately been about making me look good. If Albert had discovered the kiting, I wondered if I would have time to get investors to cover the loss, like I did in Oxford.

I stepped onto an escalator that led to the second-floor mezzanine, the main bank floor. As the mechanical stairs lifted me upward I saw Albert waiting for me at the top. He directed me to the conference room where I'd been many times before for meetings about loans, equipment leases, and acquisitions.

I sat at one end of the large oval conference table. Albert sat at the other, as if he wanted to be as far away from me as possible. A woman I had never seen sat to his left. He looked down at the table as he told me that the FDIC had been at the bank doing an audit. He stammered about some procedures the bank went through before the auditors arrived. Then he looked up. "It's come to our attention that you've been kiting checks."

I felt the blood rush into my face. I felt hot all over. But I acted as if I didn't know what that meant.

"Best we can determine," he said, "the checks total somewhere around a million dollars."

I wanted to correct him, but I thought it better to be perceived as stupid than criminal. I looked down at the table. I was ashamed.

Albert read me the bank's policy. They would not accept any deposits that were not in cash. They would notify the FBI. They would notify the other banks on the coast. They would not entertain any loans. They would begin foreclosure on any liens on my assets. Then, he added that my uncle Knox, the bank's primary attorney, had been made aware of my actions.

As we left the conference room, I told Albert I was sure I could cover the overdraft.

Looking at the floor and turning away from me, he said, "I hope you're right."

As I rode the escalator down to street level, I considered what I could do. And for the first time in years, I didn't know. I was certain of only one thing. That nothing, ever again, would be the same.

CHAPTER 37

On a crisp fall day, bundled in a heavy jacket, I waited in line behind four other inmates for the pay phone. Linda and the kids had
returned to New Orleans from their trip to Oxford and I was anxious
to hear about it. Linda accepted the collect call and spoke longingly
of the simple life a small town afforded. She had so many old friends
in Oxford. And her family was there.

"I'm moving back," she said.

"What?"

"I'm moving to Oxford."

I understood her impulse. She would be near friends, in a safe
town, out of the French Quarter, away from my mother. Her family
could help with Neil and Maggie.

I worried that a twelve-hour round-trip would keep Linda and
our kids from visiting me. A move to Oxford didn't make sense—the
kids' school tuition in New Orleans had been paid; Linda was living
rent-free. Another move would be disruptive for Neil and Maggie.
Not to mention the obstacles I would face starting over in a town
where I had left such a mess.

"I can't move back there," I told Linda.

She didn't answer.

"Why Oxford?" I asked. "What about some other small town?"

There was a long silence. Then Linda said, in a firm voice, "I'm
filing for divorce."

CHAPTER 38

For all that I had done wrong, one part of my life had been uncomplicated and good: my life with my children. On any given morning, I would rouse from sleep before Linda. On some days, so would Neil and Maggie. To see them peek around the doorway wearing their footie pajamas was a treasure. Their hair matted from sleep, breath still sweet. I would put them on my lap to ask what I could fix for breakfast—cereal or oatmeal or their favorite, strawberry Toaster Strudel.

In our evening routine, Linda and I took turns bathing Neil and Maggie. On my nights, I used too much bubble bath, splashed too much water on the floor, and sometimes forgot to wash behind their ears; but to hold them close after a bath, to smell the scent of shampoo and powder was an escape from the frantic world I had built for myself.

On weekend mornings, when I didn't rush to the office, I got down on all fours and pretended to be a horse. When my knees or back wouldn't take it anymore, we would make a pallet in front of the television—a stack of quilts and blankets so thick it felt like a pillow—and watch their favorite Saturday morning shows. During commercials, I would balance them atop my feet so they could fly like Superman, or place them horizontally across my chest and use them like weights on a bench press, or I might stand and grab their hands while they walked up my legs and torso, then flip them backward to complete a "skin the cat." Maggie called it "cat scan."

In nice weather, we would go to Second Street Park to swing and slide and spin on the merry-go-round. For a special treat, we would drive to Fun Time USA, an old-time amusement park with rides renovated from the 1960s.

At bedtime, I would read them *Corduroy*, the tale of a magical shopping trip, *The Cat in the Hat*, *Goodnight Moon*, or *Where the Wild Things Are*. When sleep escaped them, I would read the books over and over. Other times, lying in a tiny twin bed with my child, exhausted from my day at work, I would fall asleep midsentence, still holding the book above my face. Maggie or Neil would yell, "Daddy!" to wake me. And I would read on.

On nights when reading did not put them to sleep, I sat in the rocking chair that belonged to Linda's family. I would hold them on my shoulder, rub their backs, and hum "Amazing Grace" or "Rock-a-Bye, Baby" until sleep consumed us both. When Maggie was an infant, nursing from Linda's breast and sleeping in our bed, I would put her on my stomach and rest my hands at her side. Her tiny head on my chest, our breathing synchronized, the warmth from my body like a heated blanket, Maggie would fall into a deep sleep.

Linda's words—*I am filing for divorce*—stopped me cold. For the first time, I faced an unthinkable loss. My children. I looked for an abandoned hallway. A corner in the library. An empty television room. But inmates were everywhere. I couldn't catch my breath. Air didn't go deep enough. My hands trembled. I felt nauseated. I needed to cry. But I couldn't let anyone see me. Not a guard or an inmate or a leprosy patient.

I sat on a bench in the corner of the inmate courtyard. Slumped over, I could feel my heart pound. That's where Link saw me.

"Look at Clark Kent all sad and shit!" Link yelled, hoping to get the attention of some of his friends. "You step in some kryptonite?!"

I was dizzy. I wanted him to go away. "My wife is leaving me," I said, hoping he would take the hint.

"Goddamn!" Link said, laughing, like he thought this was funny. "You been lying like a motherfucker, you lost two million dollars, and your ass is in jail—what the fuck you think she gonna do!?"

• • •

That night, I fought away the tears. I didn't want Doc to see me break down. I lay awake, the light from the hallway shining directly on my face. I didn't cover my eyes with my forearm, like most nights. The fluorescent light kept me up until almost 2:00 A.M. Then the nightmare came.

I walked across a swinging rope bridge. Maggie walked in front of me; Neil behind. In the dream, I warned them both to hold tight to the guide rope. Suddenly, a wooden slat broke under Maggie. She fell through and grabbed onto a piece of dangling thread. I screamed and dropped onto my stomach. I crawled toward her when I heard the boards break underneath Little Neil. I turned to see his arms wrapped around a creaking wooden slat. Little Neil and Maggie screamed for me to help. I stretched out my arm toward Maggie, but I couldn't reach her. She looked me directly in the eye. "Help me, Daddy," she cried. I moved my body as far off the planks as I could without falling. As I reached out for her hand, everything went dark. I was completely blind. I waved my arm wildly, trying to find Maggie's hands. I yelled for Neil to hold on. Then, in the dark, I heard the sounds of their screams falling far, far away.

I sat up in bed and yelled.

"Jesus Christ!" Doc said, sitting up. "What the fuck is going on?"

I put my hands over my mouth.

"Go back to sleep," Doc said.

I was afraid to go back to sleep. Finally, 4:00 A.M. came, and I went to work in the cafeteria. Chase and Lonnie didn't report to work until breakfast time. Link, Jefferson, and other early-shift inmates settled into a deep sleep inside the cooler. I went to the empty patient dining room. Ella wouldn't arrive until 5:00 A.M.

At first the tears came slowly. Soon all that I had held inside gave way. I cried until I choked, until watery, clear mucus ran down to my chin. I cried until the tears wouldn't come anymore, until I gasped for air and spasmed with each dry heave. I cried until I was exhausted and listless.

I used the leprosy cafeteria restroom, off-limits to inmates, to wash my face. My cheeks were covered with red splotches, like my skin was allergic to my own tears. My eyes were puffy and bloodshot. In the mirror, I stared at a face I didn't recognize. The face of a man who had lost everything.

When I returned to the cafeteria, Ella was pouring a cup of coffee. I didn't want her to see me like this, so I hurried over to the menu board, my back to her.

As I wrote the day's menu, I heard the clanks of the chains turning on her wheelchair. "Hey, boy," she said. I waved over my shoulder and went back to the board.

"You all right?"

I turned around, and Ella saw my face.

"Sit yourself down," Ella said, touching the table.

I slid down into a chair next to her and told her about Linda leaving me. I told her I would never again live in the same house with my children. I stretched my arms on the top of the table and put my head down. Ella turned her wheelchair to the side and reached for my hand. Her palm felt cool and smooth. Her skin smelled like flowers. For nearly seventy years, Ella had suffered, and seen, heartbreak beyond anything I could imagine. She had been torn from her family and imprisoned as a child, but she offered me comfort without judgment or comparison. Ella sat with me, her hand on mine, in silent vigil.

CHAPTER 39

I went to my room, crawled into bed, and pulled the covers over my head. I told the guards I was sick. I avoided everyone. I slept through meals and woke periodically in the night, disoriented, forgetting for a moment why I couldn't breathe. I lost track of time. When I couldn't sleep anymore, I sat up in bed. My frantic attempts to think of a solution to keep my children quickly gave way to crushing anxiety. For the first time, I faced losing the one thing I could not bear to lose. And no amount of charm or logic or persuasion or money would buy me out of it.

I lay awake, staring at the ceiling, and remembered a day in May 1992, three weeks after the banks had frozen my accounts. We were living in a three-story, four-thousand-square-foot house with an empty refrigerator. And I was out of cash.

My family, and Linda's, had already spent thousands investing in my business and retaining lawyers. I could not ask anyone for more. I saw the panic on Linda's face.

The next day, when I went to Hancock Bank, I kept my head down, hoping no one would notice me. I checked the contents of our safe deposit box. An old Garcia & Vega cigar box held the silver coins my grandfather had given me on each of my birthdays, as well as about a dozen U.S. savings bonds held together by a rubber band. Two bonds had been gifts to me during my childhood; the others belonged to Neil and Maggie. On the envelopes were handwritten notes: *From Grandpa Burrell on Your Birth*; *From Granddaddy Ron*; *Happy Birthday, Love, Pappy*.

The denominations ranged from $25 to $50, and although the bonds were made out to my children, I was also listed. As guardian. The person who watches over and cares for their assets, who protects them from mismanagement.

I put the bonds in my back pocket and returned the box to its place in the vault. At the teller window, I endorsed the back of each one. The woman behind the window recognized me. She kept her eyes down. I slid the bonds toward her. "I'll be right back," she said, not looking me in the eyes. She gathered them together and disappeared.

I wanted to explain to Neil and Maggie what I was doing, but I didn't know how to ask for their permission without shaking their sense of security. How could I tell them we were out of money? Out of food? How would I ever tell them I might go to prison?

The teller counted out $240 and change. I was ashamed to use their money. And I had expected more.

"Cashed before maturity," she said. Her eyes said, *You should be ashamed of yourself.* She had no idea.

With my children's money, I went to Delchamps. The smell of sour milk in the dairy section made me nauseated. I looked for the least expensive items. The ones that would stretch the farthest. Pasta, peanut butter, crackers. For the first time in a decade I paid attention to price and was careful not to overspend. I bought special treats for Neil and Maggie. Fruit Roll-Ups, Little Debbie Zebra Cakes, ice cream sandwiches.

At home, I put the groceries on the counter. Linda saw the junk food I'd bought for the kids and looked at me like I was an idiot. I shook my head and turned away.

Neil and Maggie were on the living room floor in their pajamas. I sat down in the leather armchair and held out two Fruit Roll-Ups. They jumped onto my lap and put their arms around my neck. Their hair was still wet from their baths, and I could smell the baby shampoo. They nestled into my arms. They were happy. Their father had brought them candy. They felt safe and loved and secure. Secure because they knew I would never do anything to hurt them.

The Catholic church at Carville.

CHAPTER 40

On a Wednesday afternoon, after days of crippling despair, I climbed
out of my bed. I stood under the shower until my skin turned a deep
red, hoping the hot water would wash away the pain. Back in my
prison room, towel wrapped around my waist, I leaned on the sink
and looked into the mirror. I couldn't stand the sight of my own reflection. I needed help, but I didn't know where to turn.

I dressed and went to the Wednesday night Catholic church service.
The stained-glass windows of the church, so brilliant in the sunlight,
were dark. The altar was gently illuminated by candlelight. The service
was small. A couple dozen inmates sat in the middle wing, along with
Sister Margie. Five or six leprosy patients, the most devout Catholics,
were scattered around their wing. Rosary beads in hand, the patients
chanted Hail Mary and Ave Maria intermittently. Deep in prayer,
concentrating on the great Christian mysteries, they stopped to kiss
the crucifix dangling from the bottom of the string of beads, only
to begin the ritual again. For nearly one hundred years, the leprosy
patients at Carville had turned to the Catholic church for comfort.
Counting rosary beads with numb fingers, listening to passages about
the unclean, and praying that they too would be healed.

I sat between two Mexican inmates who didn't speak English, but
I watched the leprosy patients to my left. In medieval times leprosy
patients had been banished from traditional churches. Sanctuaries
were built with a "leper's squint," a narrow opening carved into the
side of the church building where the afflicted could get a glimpse of
heaven without endangering the congregation.

Father Reynolds stood before us. He began in his quiet, unassuming voice: "We are told to believe in ourselves," he said. "But I'm not sure that's what we are meant to do."

Father Reynolds had chosen that evening to talk about pride—the foundation from which all other sins arose. Pride, he said, was as an excessive belief in our own abilities. Lost in our own pride, he explained, we are unable to recognize grace.

I felt as if he were speaking directly to me.

Early in my career, when I launched my small paper in Oxford, I simply wanted to tell a good story, serve the town's need for a legitimate journalistic voice, and support my family. But along the way, good motives took a second chair to ambition and the accolades that came with success. I convinced myself that the good I was doing justified bending the rules. And I never seemed to suffer any serious consequences. My pride spread like a cancer.

I wrote editorials proclaiming to be a watchdog for the townspeople of Oxford. I reported on crime, corruption, and conflicts of interest. Then on some days, I would write myself a bad check and deposit it in my corporate account to create a temporary balance.

When my scheme was exposed, I convinced honest men to invest more money. In short order I'd run through it. Even the threat of an FBI investigation and the shame of bankruptcy did not curtail my ambition. Instead, I abandoned my dream of journalism for a business designed to make vast sums of money.

When I launched *Coast Magazine*, I was careful to feature only the beautiful people, places, and things in my hometown. We published positive stories about those in power. Huge amounts of money flowed through the business. But I was not satisfied. When I embarked on the five-year plan to build a publishing empire, expenses soared, and the magazine income could not support my dreams. To achieve my goals, I sold my invoices to third parties, a technique called *factoring* that enhanced cash flow. And I convinced small investors to place their trust in me.

And when that didn't cover expenses, I fell back on my old technique of kiting checks. It was my own secret energy pill. It set me apart. I succeeded where others failed. CEOs slapped me on the back. Restaurateurs refused to let me pay for meals. Junior Leaguers scrambled to make it onto the pages of my magazines. My system of kiting and refinancing and factoring and more refinancing created access to cash and an image of great business acumen. It worked—and made me look successful.

To transfer hundreds of thousands of dollars between two banks across the street from each other depended on perception. The perception that my company was flush with cash, that the large checks written to my own accounts were simply transfers, that I had nothing to hide.

I didn't save money. I spent it to impress. For me, money and image had become inseparable. Kiting checks afforded me the freedom to spend and buy and pursue wild dreams, and bring everyone else along for the ride. And I never had to wait.

I consciously acted in a manner to appear trustworthy. I looked and dressed the part. I went to church. I volunteered my time with charities. I proclaimed to be the journalist who would watch over criminals and politicians and casino owners. People believed I was honorable.

By the time I was thirty-one, ambition had become the driving force in my life. Even worse, I fell prey to my own mirage. Privately, I envisioned the figure I would become—owner of a huge network of city magazines, editor of a daily paper, holder of innumerable civic awards, owner of a fabulous yacht, and, of course, philanthropist. With these images fixed in my mind, I was able to overlook what I did to get there.

But the prospect of losing my children had stripped away every pretense. It did what bankruptcy, public humiliation, and imprisonment had not done. I could no longer stomach my own lies and delusions. For the first time, I felt the full weight of my crimes.

I had cost bankers who trusted me more than a million dollars. I left thirty loyal employees without any income. I put small-business owners in a deep hole. I lost most of my mother's retirement fund, money she had invested in my business. I had disappointed my friends and family. I had put my uncle Knox, Hancock Bank's lawyer, in a terrible spot. And I had allowed a woman, a single mother who couldn't afford to lose her investment, to put her money into my company. A year later, she and her two children were evicted from their home. I betrayed Linda and left her in debt, dependent on others, drowning in the shame of all my secrets. And I left Neil and Maggie, the most important people in my life, without a father in their home.

Even incarcerated, when I should have been most humble and reflective, I held on tight to my vanity. I wanted my shirts pressed; I hoarded scent strips to smell good; and I imagined myself winning a press club award before I'd done a moment's work. When I should have been trying to change, I grasped on to the image I'd held so dear. And though I had publicly acknowledged some of the bad things I had done, I had never taken an objective look at the person I had become.

Finally, in a sanctuary for outcasts, I understood the truth. Surrounded by men and women who could not hide their disfigurement, I could see my own.

PART IV

Winter

My mother and me (in kilt) in Scotland, 1969.

PART IV

Winter

My mother and me (in kilt) in Scotland, 1969.

CHAPTER 41

I stood behind the barricade and waited for a guard to escort me to the visiting room where my mother was waiting for me. She had made the trip alone and, as usual, arrived forty-five minutes early.

During my first year of junior high, my parents divorced. I lived with my mother and my three younger siblings. My mother, perhaps to counteract any ill effects of divorce, reminded me almost daily that I had been chosen for an extraordinary path. I started to believe her. Not just that I could make a difference, but that I was special and had been called to share my gifts with the world. She believed her children's skills should be showcased at every opportunity. She registered me, as the eldest son, for races and contests and tournaments. A master at bolstering my self-esteem, she often reminded me of the meaning of my name. She would say the words like she was sharing with me my own destiny, "Neil *means* champion." I believed what my mother told me. And I was certain we would make the world a better place.

Mom started an alternative school for juvenile delinquents in Gulfport. She spent her days giving them the attention and praise and hugs they had missed in their homes. She lauded their talents, whether those came in the form of suggestive dance, graffiti art, or a knack for breaking into locked cars to retrieve keys.

To some, my mother was a saint; to others an enigma. She was on her fourth marriage; she held three graduate degrees; she had lived in twenty-seven different houses during the last three decades and had held no fewer than seventeen jobs. She had launched two magazines, founded three schools, self-published five books, served on the boards of four corporations. She had started a restaurant, a dress shop, a riding stable, a camp for disadvantaged kids, a nonprofit education company, and a low-income housing project. She had a house in Gulfport, an apartment in New Orleans, and a husband in Oxford. She took risks. She relished the unknown. She loved the limelight. She stood fearless in the face of change and approached her own life as if it were a thrill ride. And I had always wanted to emulate her.

When I arrived in the visiting room, I saw Mom waiting at a table in the back corner. We hugged, and she said, "How are you, baby?" But she didn't need to ask. She knew I was in trouble again. That's why she was here.

The last time I was alone with my mother was April 9, 1992, about an hour after the banks had closed my accounts. When I arrived at her house, she tried to hide her dread, but she knew something was wrong. I wouldn't have been at her house, on her back porch, in the middle of a workday if the news were good. Mom lived in a house that had been handed down through three generations of her family. It was once my great-grandmother Floy's home. Floy was a teacher and missionary. In 1903, she moved to the Philippines. She educated the islanders about math and literature and God. Floy had retired by the time Mom was born, so Mom became Floy's student. She instilled in my mother a sense of service and selflessness. In the 1930s, Floy invited people of color to sit and eat at her dining table when integration of any sort in Mississippi was taboo. She fed hoboes who jumped off the train between New Orleans and Mobile. And she passed along lessons to my mother: "If you have an extra dollar," Floy would tell her, "give fifty cents to someone who needs it more than you—with the remainder, buy a book." Floy's rules for living were passed along to my mother on the very spot where we sat on that cool Thursday morning.

With the breeze from the Gulf of Mexico gently rocking her porch swing, I told my mother I was over $2 million in debt. I told her I had no idea how I would repay the $200,000 she had invested in my company. I told her the FBI would be investigating to determine if I had violated any laws. My mother pressed her lips together and started to cry. She hugged me, fighting tears, and said we would get through anything as long as we stuck together. I left her sitting in a chair on the porch. She put her head in her hands and let the tears flow. She was the strongest woman I knew, but this was too much. She cried for Linda. She cried for Little Neil and Maggie. She cried for me, her firstborn son, who might face prison. She cried because she had no more money to give me. She cried for all that we were about to lose.

Mom reached across the visiting room table and held my hand. Mom, at fifty-two, was still a striking woman, tall and fit, with auburn hair. She was strong and generous. She never once mentioned the money she lost investing in my company. And she still seemed to believe all those things she told me when I was a child. In spite of everything, she still believed in me, and that I would do great things.

"I talked to Linda," she said. "I'm going to bring the kids to visit as much as I can."

Stability wasn't Mom's strong suit, but no one was better in a crisis. From Oxford to Carville was a twelve-hour round-trip. But Mom lived in Gulfport. The drive to collect the kids, bring them to Carville, and then return home made it a twenty-four-hour ordeal in a single weekend.

Mom took a deep breath and rubbed her hands over the linoleum table like she might be sweeping away crumbs. "Baby," she said, "what are your plans?"

"I have no idea," I said. My postprison career plan had disappeared. Linda and I had discussed launching a small publishing venture. She would be the front person, we had decided, and I would work behind the scenes. But now, with an impending divorce, I didn't have a plan.

"Do you know where you'll live?" she asked.

Memphis seemed like a logical choice. It was the largest city near Oxford. I would live less than an hour away from Neil and Maggie. Surely I could find a job there.

Mom looked away like she was disappointed. "You need to think long and hard about that," she said. She put her hands together and leaned on her elbows. "If you live in Oxford," she said, "you could see them every day."

I could list a thousand reasons to not move back to Oxford. I reminded Mom that no good jobs existed in that small town, especially for me—an ex-convict who, five years earlier, had alienated so many in Oxford, ended up in bankruptcy, and moved away disgraced. Not to mention Linda, who would abhor the thought of my following her.

Mom swallowed. "You either live in the same town with your children . . . or you don't. There is no in-between." She talked about the reality of a seventy-five-mile separation. "If you build a life in Memphis, you will grow apart from your children. It breaks my heart to think about your living in a place where Neil and Maggie don't live."

I told her I would give it some thought over the next few days. She said she would bring Neil and Maggie to visit often, and she would make sure they were here for Kids' Day—a Saturday when inmates' kids would be allowed inside the prison for the entire day.

When visiting hours ended we hugged and said good-bye. As she left, a middle-aged inmate asked if I would make an introduction the next time she visited.

Maggie, me, Little Neil, and inmate Steve Read (in clown suit) (left to right) *during Kids' Day celebration.*

CHAPTER 42

On a Saturday morning in December, I waited with about forty other inmates for a guard to escort us to the ballroom for Kids' Day. I was almost giddy about spending the day with Neil and Maggie. For the first time they would be allowed beyond the visiting room. We would be able to spend time together inside the prison, and they would finally see where I slept and spent my days.

The guard who escorted us was overly polite. He said more than sixty children were waiting for us, their inmate fathers. A dozen guards and a handful of inmates had volunteered to run the activities for the kids.

When I walked into the ballroom, I saw Neil and Maggie standing with some of the other children. Maggie wore a denim jumper and a bright red Christmas sweater. Neil was dressed in a hooded sweatshirt and sweatpants. When they saw me, they both ran toward me and jumped into my arms.

Steve Read was one of the inmate volunteers. He was dressed in a clown suit. He set up a station where he made balloon figures for the kids. He made reindeer antlers for Little Neil and a sword for Maggie.

A female guard handled face painting. She painted Maggie's face like a cat. She painted Neil's nose red and gave him a huge circus-clown smile.

An inmate band of four Mexicans played the same song over and over. Two counterfeiters put on a puppet show for the kids. Even in prison, there was a carnival atmosphere. Throughout the morning the guards and inmates organized cake walks, a beanbag toss, and dizzy-izzy contests. Bean, the chubby Mexican inmate, took photographs of the inmates and children.

All the guards were exceptionally nice on this day. And I appreciated it. In fact, they served us. As we sat at tables with our children, guards brought us platters of hamburgers and hot dogs. They poured lemonade and mint-flavored ice tea from a glass pitcher.

After lunch, the kids were told to rest before the outside games began. Neil and Maggie wandered around the ballroom. Neil found one of the patients' bingo cards and asked if we could play. It was a specially designed card where chips weren't necessary. Patients who had lost feeling in their fingers couldn't handle the small chips. I imagined Ella had used this card on occasion. She loved bingo.

Then Neil saw the grand piano. He had just begun to have an interest in playing. I asked a guard if it was all right for the children to play it. She said yes. It was a full-sized grand piano, but it was in terrible condition. The ivory had fallen off some of the most-often-used keys in the center. As Neil and Maggie put their small fingers on the piano keys, I thought about how many leprosy patients might have placed their fingers in the same place. When the piano was first purchased, before there were medicines to stop the progression of the disease and when finger absorption was commonplace, sometimes two patients would learn to play a song written for a single pianist. They learned to play the songs as duets so they would have enough fingers between them to play all the notes.

Rain started to fall, and the outside activities were postponed. But the guards had a backup plan. They had rented a movie, *Free Willy*. As the guard slid the VHS tape into the deck, the kids took their places on the floor around the large television. An inmate whispered in my ear. "What kind of fucked-up people would show a movie about an imprisoned whale to a bunch of kids visiting their dads in prison?"

Steve Read, still in his clown costume, added, "*Free Daddy* must have already been checked out."

Two hours later, when Willy jumped over the barrier and swam away, the rain stopped, and the kids were allowed thirty minutes inside the inmate courtyard—and then, for a few minutes, inside our dorms. This is what Neil and Maggie had been waiting for. They didn't understand why, during their visits, they couldn't play in the inmate recreation yard. Maggie, particularly, didn't understand why she couldn't come to my room. Today, as I promised, they would play on our basketball court, run around our track, and try out my prison bunk.

We walked downstairs past the post office and the leprosy patient canteen. Ella was just making her last run of the day. I stopped to introduce Neil and Maggie to her. She smiled and held out her hand, but they had no interest in talking. They were impatient and didn't want to miss anything, especially the slam dunk promised by the six-feet, seven-inch inmate named Slim. Neil and Maggie ran toward the inmate side. I shrugged. Ella understood. For her to watch sixty young children running through the colony corridors must have been surreal. But she seemed happy to be in the middle of it. I left her sitting there like a boulder in the middle of a stream of children.

CHAPTER 43

Outside, in the inmate courtyard, Slim waited for the children to join him on the basketball court. He performed a series of basketball dunks. He palmed the basketball and then let the children compare their hands to his. Maggie ran through the sand on the volleyball court. Then, we bounced a ball against the wall of the outdoor hand-ball court.

A guard warned us that we had fifteen more minutes before the kids had to leave. I led Neil and Maggie up the stairs to my room. Doc had gone to the library to give me some privacy with them. The kids were excited to see my room, even though there wasn't much to show. I opened the closet door to show them the huge piles of Doc's medical journals. I sat at the desk to demonstrate where I sat at night to write them letters. I put them both on my bunk and opened my locker door to show them how I had a perfect view of their pictures.

I pulled a chair over to my locker so Little Neil could stand on it and explore its contents.

I put both my pillows against the concrete wall, took off Maggie's sandy shoes, and propped her up against the pillows. She put her hands behind her head, stretched out her legs, and closed her eyes.

Neil rummaged through my locker, hoping to find some undiscovered possession. I remember doing the same thing to my father's armoire. Neil discovered the chocolate cookies and fudge that Sergio, a Cuban inmate, had made. Maggie, her faced still painted like a cat, stretched out on my bed. She moved her tiny feet across my gray wool bedspread. "Can we spend the night, Daddy?" Maggie never had understood why she couldn't spend the night.

"I wish you could, sweetheart. I wish you could."

Maggie let out a sigh, the kind I made in my own bed after an exhausting day. She looked content, as if she never wanted to leave this spot.

Before coming to Carville, I had worked to pay for vacations and expensive toys. I thought a fabulous home and fast boats would make us a happy family. But Neil and Maggie felt completely at home in a tiny room that was designed for leprosy patients and now housed federal convicts.

As I watched my son dig through my paltry belongings, excited to discover any new treasure, no matter how small, and Maggie, completely content, resting on my prison cot, I made myself a promise. I would never let anything prevent the three of us from having a life together. A simple life with time to play.

Chains rattled as a guard walked toward my room. Our time was up. I held hands with Maggie and Neil as we walked the corridor where leprosy patients had walked before us. Mom was waiting in the visiting room. She would drive the kids back home. One of the happiest days I could remember was coming to an end.

But it seemed like the beginning of something wonderful.

CHAPTER 44

A few weeks later, Steve Read finally invited me to participate in his Friday night Monopoly game. "There are two house rules," he said. "I'm the sports car. And there is no limit to how many hotels you can put on a property. Oh, and don't even think about buying Board-walk."

Steve's room had been totally rearranged for the game. He had finagled a card table from the recreation department. The cash and properties had been spread out on a side table. Steve had everything planned. He pointed to me. "Hey, you check-kiting, overdraft-king motherfucker, you can be the banker." Then he pointed to Gary. "And you, mortgage-fraud scum-broker, you are in charge of the property deeds."

"What are you going to do?" I asked.

"Win," he said, "and keep the game moving."

Steve was in his element. High finance, wheeling and dealing—at least as close as you can get to that in prison. He was on an adrena-line high. He moved the game pieces for all the players. If we didn't grab the dice in a timely manner, he rolled for us. He couldn't wait for his own turn to come around. He was impatient with us. Even when he did let us roll, he calculated where the game piece should be on the board, grabbed it, and moved it to its appropriate spot without counting.

Steve had something clever to say about every move. When Gary read aloud the "Go to Jail. Go Directly to Jail" card, Steve asked, "Is anyone else having déjà vu?" And when he drew his own Get-Out-of-

Jail-Free card, Steve said, "I wonder if the warden would accept this."

In short order, Steve took a lead in the game. We were all out of our league. He had more money than the rest of us. He had purchased Boardwalk and Park Place, and he gloated about the ensuing slaughter. Soon, Gary and the other player were eliminated. I was once again on the brink of bankruptcy.

Then Mr. Levin walked into the room.

"Sorry, Levin, no Jews allowed," Steve yelled. "He'll try to take over the bank!"

Mr. Levin and I were friends. He prepared five hundred slices of toast each morning in the cafeteria while I wrote the menu boards and prepared the garnish. He was one of the few inmates who didn't spend his mornings in the cooler. On the outside, Levin was an attorney who represented New Orleans' top Mafia boss. I told Mr. Levin I could use some help. I figured a guy who helped Carlos Marcello build a legal fortune in real estate and investments would come in handy in a Monopoly game, especially considering my recent business record.

He took a seat where Gary had been. With each of my moves, I took Levin's recommendations. He suggested I buy railroads.

Steve laughed at the strategy. "What's with you and the railroads, Levin?"

"My grandfather worked for the railroads," Levin said. "They're good properties."

With Mr. Levin's help, I soon acquired all the railroads; Steve purchased hotels for Boardwalk and Park Place. Rent would be $3,000. If I landed on either property, I was out. As Levin and I discussed my options for hotels on my cheap properties, Steve was hyperactive. His hands were moving so fast across the board, rolling for both of us, moving our tokens forward, reshuffling the cards and money on the board, I couldn't keep track of all he was doing, but Mr. Levin did. He corrected Steve a couple of times on his counting and then caught him stacking the deck. Steve had shuffled the Chance cards so that the top card was the "Take a Walk on the Boardwalk." Levin cleared his throat and held the card in the air. Earlier in the game, it had been put on the bottom of the stack. Steve had placed it back on top. Steve knew if I landed on Chance, the game would be over.

Levin reshuffled the cards, put them back in a random order and suddenly took a greater interest in my success. He kept track of my money, paid my rents and fines, collected when Steve owed me, and continued to advise me on houses and hotels. Levin's advice was flawless. I accumulated more property and more money. Steve had mortgaged everything to put eight hotels on Boardwalk. He was down to $180 in cash when he landed on a railroad. Rent: $200. With Levin's advice, I won.

Steve stormed out of the room. I thanked Mr. Levin.

"Call me anytime," he said.

CHAPTER 45

The next morning, while transcribing the menu board in the patient cafeteria, I thought about Mr. Levin. As ridiculous as it was to put so much stock into a board game, I had beaten Steve Read, a financial shyster. Considering my recent history with money, it felt like an accomplishment. All I had to do was accept help, albeit from a mob lawyer. For a moment I imagined how Carlos Marcello must have felt with so many people like Mr. Levin watching his back, protecting him, silently ensuring the success of his ventures. At once, I felt guilty . . . and oddly privileged.

On the outside, I was reluctant to ask for help, unwilling to share credit, adverse to guidance. I wondered how my life might have turned out differently if I'd been more willing to accept the help of others.

I heard Ella's wheelchair behind me. I called Mr. Levin over from the inmate cafeteria and introduced him to Ella. I recounted our Monopoly game to her, including Mr. Levin's perfect strategy and advice. Mr. Levin smiled and pulled a Chance card from his front pocket—the "Take a Walk on the Boardwalk." He had never put it back in the deck.

Mr. Levin had cheated on my behalf. And all of a sudden I didn't feel so great.

"You would have won anyway," Mr. Levin said. "Just a little insurance." He smiled, put the card back in his pocket, and went back to making toast for the inmates.

"How you doin', boy?" Ella asked.

I told her I was confused. Confused about so many things, but especially about where to live.

"There's no place like home," she said.

I usually didn't answer when Ella chanted her favorite phrase but, today, it was pertinent. "I'm not sure where mine is," I said. I tried to explain to Ella the complexity of moving back to Oxford. But to keep my resolve to live a simple life with my children, Oxford seemed the only choice. "My mother thinks I should move to Oxford."

"Sound like she know somethin'," Ella said.

Ella was right. Mom did know something. She spoke from experience. When I was fourteen, she moved away with her new husband. I stayed in Gulfport to live with my father. She had missed years of waking up with me in the same household. She had missed my games and performances. She missed holidays. She had missed my life as a teenager.

Mom had mentioned regrets in passing, but the last time we spoke she was saying something more.

"Maybe she does know something."

"Childrens need they mommas *and* they daddies," Ella said.

"How can I face the people of Oxford? What will people think?" I said.

"What peoples think," Ella said, "ain't none of your business."

That night, in bed, I pondered this novel idea—to act without seeking praise from others. A good portion of my adult life had been spent daydreaming about what others thought of me. I imagined and reimagined accolades, awards, trophies, applause. *Just wait until they see this!* I would say to myself, not even sure who "they" were.

Journalism had been the perfect profession to spread the good news of my accomplishments. More than sixty thousand households—every neighbor, friend, and relative—received a monthly sampling of my works bound in the finest paper money could buy. People stopped me on the street to talk about a never-before-published photograph I had discovered, or a thought-provoking editorial I had penned. And I was more than happy to stop and elaborate. At times, it made me dizzy. I felt like I was fulfilling a destiny.

Now, this hunt for adoration felt demeaning. Linda's news had been the catalyst that convinced me I needed to change. But I had no idea how to replace a drive to impress that had become second nature. Since childhood I had been imagining myself on the pages of *The Guinness Book of World Records*, or inviting an audience to watch me fly from a tree house roof.

Ella couldn't afford to imagine what people were thinking. If she did, she might never leave her room. Her approach was so simple: *Here I am.* She rolled into a crowd of inmates, made eye contact, and expected us to treat her with kindness.

I had fewer than five months before I would be set free. And to be like Ella, I had to unlearn a lifetime of habits.

CHAPTER 46

"Have you seen this?" Doc asked. He handed me a memorandum from the warden. According to new rules, inmates could possess no more than 5 books, 25 magazines, 60 cigars, and 288 cigarettes.

"What's the rationale?" I asked.

"You're assuming these idiots are rational," he said. "They say it's a fire hazard." Doc had accumulated more than 300 medical journals.

"What are you going to do?"

"Ship them to my girlfriend," Doc said. "Should make her day."

I opened my locker, removed about a dozen books my mother had mailed me, and spread them out on my bed. I wasn't a big fan of self-help books, but Mom had a bookcase full of them ready to share with the suddenly self-aware. I picked five books with titles that sounded like they might be the most useful, put them in a stack, and grabbed my pad and pencil.

"You're not going to read that crap, are you?" Doc asked.

I put the books under my arm and walked to the cafeteria. After I finished the menu board for the leprosy patients, I flipped through the first book, *Tough Times Never Last, but Tough People Do*. The back cover read: *"Make your dreams come true! Dr. Schuller shows you how to build a positive self-image."*

I skimmed the book. A few emboldened phrases jumped off the page: *"Take Charge and Take Control. Count to Ten and Win! Dare to Take a Risk. Only a Person Who Risks Is Free."*

I put the book down and stared out at the leprosy patient court-yard. The last thing I needed was a new voice telling me to take risks.

I made a note to send *Tough Times Don't Last, but Tough People Do* back to my mother.

Ella rolled into the cafeteria. "What you readin'?"

I held the books up for her to see—*Divorce Without Victims, Untwisting Twisted Relationships, Homecoming*, and, of course, *Pleasing You Is Destroying Me*. "They're self-help books," I said.

"They helpin'?" Ella asked.

I shrugged and imagined how ridiculous I must have looked, turning to paperback books to solve my problems, to a woman who had endured as much as Ella.

"Do you read much?" I asked.

"The Bible," she said. "And I looks at the paper."

Ella was certainly wise. And I valued her advice, but I suspected she might be illiterate. The leprosy patient library was well stocked and larger than the inmate library. I had passed it many times and had admired the thousands of books on the shelves. "Do you ever go to the library?" I asked.

"Nope."

"Are there are any books you want to read?" I asked. Like most journalists, I assumed the wisdom of the world was contained on the printed page. "You could bring them here in the morning, and I could read out loud."

"Ain't no need," she said. "I already knows more than I understands."

I wasn't sure what Ella meant by that. Could be brilliant. Or not. Sometimes, her words seemed out of context like they flew in from another conversation. Other times, she uttered a phrase that opened up my world. I figured it wouldn't hurt to give her a shot at my new challenge.

"How does a person change?" I asked Ella.

"Hard to do," she said. "Hard to do."

"Any ideas?"

Ella hesitated. She generally blurted out whatever popped into her mind.

"What?" I said, trying to prod her.

"Keep meddlin'."

"I don't want to interview anyone else," I said.

"Maybe," she said, "you been askin' the wrong folks."

I had talked to Father Reynolds, Reverend Ray, Sister Margie, every interesting inmate, and dozens of leprosy patients. I already had volumes of notes, but if there were a counselor or a nun or a psychologist who knew something about how to change, I suppose I needed to ask for help. And Ella obviously thought someone could offer insight.

"Who?" I asked.

Ella leaned forward and in a soft voice said, "Your own self."

I walked the track tossing Ella's words around in my head: *What people thinks about you ain't none of your business. I already knows more than I understands. Interview your own self.* As I moved in circles around the inmate courtyard, I decided to take Ella's latest advice. I would treat myself like any other interview subject. Ask hard questions. Be suspicious of motives. Look for cracks in my own story. Treat myself like a man who had every chance and ended up in prison for bank fraud.

"Hey, Clark Kent!" Link yelled. "You borin' motherfucker!" Link joined me on my walk around the track. "What you doin'?"

"Trying to figure out some things."

"Tryin' to figure out why you so white?"

I didn't answer. I stopped to make a note on my pad. Link waited. Then we walked again.

"You know all that shit I told you before?" Link said.

"What?"

"You know," Link said, "'bout Popeyes and takin' them cars."

"Yeah. I remember."

"You can't write about that shit . . . unless you give me some money."

"You need money?" I asked.

"If you put me in your story, I want a million dollars," he said, smiling.

I tried to explain the premise that if information is related directly to the journalist, without the caveat of being "off the record," the writer has every legal right to use the material.

"That some bullshit!" Link yelled. "You can write shit 'bout me without giving me a goddamn cent?!"

"Yes," I said. "But, if the information is libelous, litigation is a remedy."

"What the fuck?"

"If a reporter writes something about you that is inaccurate, or reports the story in a manner that is damaging to you, you can sue."

"Then give me some fuckin' money or I'll sue your ass."

"The stories you told me," I asked, "are they true?"

"Fuck, yeah!"

"Look, Link, I don't know if I'll ever write anything about this place." I told him I wasn't feeling much like a journalist. I couldn't very well write someone else's story when I wasn't sure of my own.

"I seen your name in that magazine," Link said.

The prison library had recently received back issues of *Louisiana Life*. I was listed as "Publisher" at the top of the masthead.

"If you put me in one of them magazines, you got to give me some motherfuckin' cash."

"I don't own those magazines anymore." I wanted to tell Link that the socialites and ladies who read *Louisiana Life* had no interest in reading about his escapades. Not to mention that the new owners would never consider publishing anything I wrote. "In fact," I said, "I'm in jail because I fell in love with those magazines." I told Link how much money I spent on the best lithography in the South, spectacular photography, shiny UV coatings for the magazine's cover, and the finest imported paper.

"Goddamn!" Link said. "You is doin' time over glossy paper!"

He had a point, but Link needn't worry about my writing about him anytime soon. The story—or exposé—I had envisioned when I arrived at the colony didn't exist anymore. The only thing I had exposed was my own shortcomings. I had uncovered nothing sensational, except, perhaps, that living with the victims of leprosy had turned out to be a strange blessing for me.

"Don't worry, I'm not writing about you anymore."

"Then why you still writin' in that notebook?"

I shrugged. Link smiled like he'd caught me in a lie. I would continue to make notes, but not for the reasons Link imagined.

The Christmas card from Oxford.

CHAPTER 47

Every day after the four o'clock stand-up count, the Dutchtown inmates gathered on the ground floor of our unit for mail call. As the guards called my name, I felt a bit guilty about the amount of mail I received. I had reached out to every remaining friend. And my mother, who had a knack for galvanizing people into action, had encouraged all her friends to write, too. I received dozens of magazines, three daily newspapers, and a regular supply of books and letters. Most, I'm sure, orchestrated by my mother's asking friends to write.

And Mom was particularly proficient, too. She sent newspaper clippings and snippets about my former interests. She wrote about the births and marriages of friends and acquaintances. She sent photos of our family. And though the photographs were, at times, exciting to receive, they also reminded me of the moments I was missing.

I received a Christmas card from Linda. The photograph, of Linda and the kids in front of the courthouse in Oxford, was stunningly beautiful. Wreaths with red bows had been hung between each white arch of the courthouse facade. The photograph was taken on a cool day in the late afternoon. Bits of sunlight illuminated the courthouse walls between the shadows cast by the leafless limbs of the oak trees. Linda, arms wrapped around the children, wore a thick white sweater with a dark peacoat and scarf. Maggie wore a red velvet dress with a lace collar. And Little Neil wore a blue blazer with a red-and-blue striped tie.

It was the saddest Christmas card I had ever received. It was my family without me.

One thing was clear. She was moving on. I put Linda's card in my locker and took off my wedding ring.

I wanted to perform some kind of ceremony. I imagined I would dig a small hole in the ground somewhere on the colony grounds. A burial. I would leave the symbol of our marriage here, where it ended, at the colony. But what if, by some miracle, we got back together? If I left the ring buried on the grounds, I'd never be able to retrieve it.

So instead, I placed my wedding band inside an envelope. I sealed it and wrote on the outside, Wedding Ring. I found an empty slot in the back of the expandable file folder Doc had given me, and I buried the envelope deep inside.

CHAPTER 48

My nightmares about the children persisted. In every dream, Neil and Maggie were falling from a building or a tall tree or a window in an apartment. The setting changed, but there were two recurring themes: I could never quite reach them, and I could not see the ground. And at the end of each dream, like the swinging bridge nightmare, my children fell into an abyss. I would sit up in bed, wailing, covered in sweat.

Doc even got used to it. He would groan and put a pillow over his head. On the nights when the dreams came to me, I would climb out of bed and walk the corridors until it was time for work.

Ella noticed the bags under my eyes.

"Nightmares," I told her. "Ever have them?"

"Musta," she said, "just don't remember." But Ella did tell me about another dream.

"I'm just a little ole thing," she said. "My momma holds me like a baby."

In the dream, Ella's mother held her against her breast, rocked her in an old wooden chair, and sang hymns. Ella got tickled. "She tryin' to get me to go to sleep, but I'm already asleep."

"I be all warm," Ella said. "I's a baby, but Momma already know I got this disease."

Ella's dream had a recurring theme, too. Her mother said the same words.

"Momma say, 'You got what Jesus talk about in the Bible. I wouldn'ta throwed you out.'" Ella smiled. "Then we both laughs, and I goes to sleep."

Listening to her describe this dream, watching her laugh, witnessing the way she held herself, I realized that, somehow, Ella had escaped the shame of leprosy.

I'd read about a brief period in medieval Europe when some Christians considered leprosy a sacred disease. Infection, among the most devout, was seen as a privilege. Being a leper, one of Christ's poor, meant a sufferer need not wait for any rapture. Resurrection occurred immediately. The belief that leprosy was a godly disease was so widespread that Lazar houses and leper colonies were like monastic retreats. One European prince proclaimed that putting his hands on an outcast, washing the open wounds on a leper's feet, would get him one step closer to heaven. And Father Damien, the Martyr of Molokai, had perpetuated that conviction. He said that if he were to contract leprosy—which, in the end, he did—he would gain a "crown of thorns."

But this view of leprosy had disappeared. Over the last five centuries, leprosy, and all the stigma that goes with the disease, found its way back into the human psyche.

But Ella carried her leprosy like a divine blessing. She had faith that she would be healed in heaven. She embraced the life she believed God had chosen for her on earth. She had transcended the stigma that crippled so many.

CHAPTER 49

Mom brought Maggie and Neil to visit as often as possible. Our hours together in the visiting room were a precious time. As we played and talked, my mother sat quietly in the corner and read. She could not have given me a greater gift than those moments with my children.

When Neil and Maggie weren't on the road to see me, I wrote letters reminding them that "Daddy's time-out" wouldn't last much longer. I wrote about the exciting things we would do together when I was released.

I created two comic strips with Neil and Maggie as the superheroes. Maggie's comic was entitled *Magalina Ballerina*. The heroine was a four-year-old girl who used ballet moves to fight crime and save her friends from danger. Neil's comic was entitled *Hoverboard Boy*. The hero, Little Neil, used his superpowers and a hoverboard—a flying skateboard like the one used in the *Back to the Future* movies—to save the world from evil.

As I was illustrating one of the comics, Steve Read looked over my shoulder and snorted. It might seem a bit strange to see a dad in prison dreaming up plots where his children fought crime, but, then again, I didn't feel like a convict when I was being a father.

"Have they captured any check kiters yet?" Steve said.

Steve could be a real ass. But he was funny, too.

When I wasn't writing to, and illustrating comics for, Neil and Maggie, I wrote to everyone I ever knew. I wrote to the victims of my crime to apologize, to make amends, if they could imagine some way for me to be helpful. I wrote to old college buddies. I wrote to old girlfriends. I wrote to former employees. I wrote to high school teachers. I wrote to other journalists and writers. I wrote to friends of my parents. I wrote to Judge Gex. I wrote to my buddy Willie Morris. The return address on my letters included my inmate number, as well as Carville's address.

Some afternoons after mail call, Link would follow me back to my room to watch me open my envelopes. At times, he asked me to read my letter out loud. He particularly enjoyed my mother's letters. He thought her expressions were hilarious.

"What'd you get today, Clark Kent?" Link asked.

I held up a copy of the *Wall Street Journal* and *USA Today*, a couple of magazines, a package from my mother, and a letter from my old friend Willie Morris. Months earlier, when I was still in the denial phase, I had written Willie and described some of the characters living in Carville. I explained my George Plimpton–esque plan, and asked if he knew any editors who might be interested in a participatory journalistic piece. Willie knew just about everyone in the literary world. He was vaguely encouraging, and he suggested that I just continue to write. Willie was too kind to point out the absurdity of pretending to be an undercover journalist during a prison term, but his tone said it all. He had lost the enthusiasm he had for me earlier in my career.

Despite the mood of the letter, I was proud that he had written me. "Willie Morris," I told Link, "is one of the great living southern writers."

"What he write?" Link asked.

"*North Toward Home. Good Old Boy. The Courting of Marcus Dupree.* He was editor of *Harper's* magazine at age thirty-one." Link didn't look remotely impressed. "The youngest ever," I added.

"Another borin'-ass, white motherfucker," he said.

I tore open the large manila envelope from my mother. It was copies from another self-help book. A note was scrawled in a corner— that I should never forget how precious I was. Mom's handwriting was terrible, primarily because she was always in a rush. She once made our school lunch sandwiches so hurriedly she left the plastic wrappers on the Kraft cheese slices.

Link sat on the side of one of the bunks. He was quiet, which was unusual.

"Anything wrong?" I asked.

Then he blurted out, "Write me a letter."

"Sure," I said.

"To my mama." Link's mother had come to the visiting room, but he had always refused to see her. He told me all she wanted to do was read the Bible to him. Eventually, she quit coming. I took out a pad and pen and told Link I would be happy to write a letter on his behalf.

"I'll write," I said. "You dictate."

"Dick what!?" he said.

"You talk and I'll write."

"Why the *fuck* didn't you say that!"

I wrote the salutation and read it, "Dear Mom." Then I asked Link what he wanted to say.

"You writin' it," he insisted. "Say whatever the fuck you want."

Link refused to offer any text, so I wrote, on his behalf and spoke aloud: *Dear Mom: This place is like a country club.* Link smiled and nodded. *I've made some friends while I've been here. The guy writing this letter for me is one of them. I call him Clark Kent. He is the whitest man you will ever meet.*

"That's good," Link said.

I continued: *I play lots of cards and dominoes. And I help leprosy patients in the cafeteria.*

I continued to write, as best I could, about Link's days—until he finally offered to help.

"Ask her this . . . ," he said, hesitating, like he might be embarrassed. "Ask her how Ashley is."

"Who's Ashley?"

"My little sister," he said. I could tell he really cared about her, and he wanted to know what she'd like for Christmas. He wanted to know what she was doing with her friends. And he warned her not to get in the car with a guy named Little Feets.

"Do you want me to ask if Ashley will come visit?"

"Yeah," Link said, pointing at the pad. "Write that down."

I finished the letter, addressed and stamped the envelope, and told Link I would mail it in the morning.

"Hey," he said, "it's OK if you write all that shit I told you."

I nodded and shrugged.

"No," he said, "I want you to."

"Maybe someday," I said.

"Don't wait too long," he said. "Life span of a nigger in my neighborhood is short."

CHAPTER 50

"Hey, Doc," I asked, interrupting his reading, "what are you going to do when you get out?"

"Leave the country," he said.

Doc rambled about the FDA's enforcement of U.S. laws and said he would never subject himself to their interpretations again.

"Where will you go?"

He thought for a moment. "Latin America might work." Doc added that Hispanic men viewed impotence as a direct reflection of their manhood. "But my heat pill has other applications."

Doc's heat pill perfectly mimicked fever. Fever was the body's natural mechanism for fighting infection and disease. Other thermogenic compounds might raise the body's temperature a degree or two, but Doc's heat pill had no limit. That was its great potential, as well as its danger. An overdose would cook a patient from the inside. But regulated and supervised by a doctor, it could generate enough heat at a cellular level to kill just about any bacteria. And no other physician on earth had as much experience with the thermogenic agent DNP as Doc. He had observed its effects in thousands of patients at his weight loss clinics. Over the years he had developed the art of dosage and frequency. And he discovered hormone supplements to reduce side effects and get the greatest results.

"You know," Doc said, nonchalantly, "the AIDS virus dies at 107 degrees." When I asked for details, he explained that the virus could live outside the body for about seven hours, but it died within minutes of exposure to 107-degree heat, and his heat pill could produce those kinds of temperatures.

"You've got to tell somebody," I said.

But Doc had no intention of telling anyone about the potential benefits of his heat pill. At least not until he was a free man. He couldn't profit from its use as long as he was incarcerated.

A few minutes later, Doc said, "It's not just AIDS." His heat pill could potentially cure Lyme disease, several forms of cancer, and just about any other infection sensitive to heat. Doc said his pill could even save lost mountain climbers. If they carried a couple of his pills, the thermogenic effects would prevent frostbite for days until rescuers arrived.

"What about leprosy?" I asked. Leprosy preferred the cool parts of the body. If Doc's pill generated heat within the cells, it would certainly kill *Mycobacterium leprae.*

"I guess it would work," Doc said. He seemed less than enthusiastic. Victims of leprosy, for the most part, lived in underdeveloped nations. It attacked the malnourished, the poor. They didn't have the resources to buy Doc's heat pill.

Sometimes, late at night, in the fuzzy state between sleep and consciousness, Doc would think aloud. In a groggy voice, he would question the logic of confinement or speculate on a new prison regulation or contemplate a recent scientific discovery.

"Doc," I asked, not knowing if he was asleep or not, "can your heat pill really cure cancer and AIDS?"

"Maybe," Doc said. "But I've been thinking a lot about growth hormones." The room was quiet for a moment, then Doc asked what I thought would be the best market for growth hormones.

The largest was obvious. "Asia," I said. "China's got a billion people."

"Yeah," Doc said, "short little fuckers over there." He pulled the gray blanket over his shoulders and turned toward the wall.

I lay in the dark a few feet away from a man who could, conceivably, cure any number of humanity's dreaded diseases. But he had no intention of sharing the cures with anyone. Doc was biding his time.

CHAPTER 51

"Ella," I asked, "do you have any children?"

"They wouldn't let me have none," she said.

"Did you want kids?"

"Wanted lots of 'em," she said, her voice trailing off. She looked down at the floor. "But they wouldn't let me have none."

Ella's smile disappeared, and I wished I could take back my question. I didn't know what to say. We fell silent.

Then Jimmy Harris called out, "Hey, young fella!" Jimmy had severe curvature of the spine, and he wore bright red suspenders that accentuated his stooped posture. He waved me over to his table.

"I'm Catholic, you know," Jimmy said. "My wife and I used the rhythm method, and it isn't very reliable." Jimmy and his wife became pregnant with a son. But rather than have the Sisters of Charity take his boy and place him in an anonymous home, Jimmy arranged for his son to live with a woman in Ville Platte, Louisiana. Two years later, when they gave birth to a girl, the same woman took the child.

"My children were raised by a wonderful woman," Jimmy added. "A saint."

"Did you get to see them?" I asked.

"See 'em all the time," he said, smiling. "They're coming with the grandchildren to pick me up this weekend."

While Jimmy talked about his children, Ella left the cafeteria. I watched as she rolled out toward the corridor. It was the only time I had ever seen her look sad.

Jimmy looked over his shoulder and whispered, "Not many people will tell you this, but in the old days they encouraged us to get an operation."

"What kind?"

"Sterilization," Jimmy said in a low voice. "They didn't force it on us, but they dangled privileges out there to encourage volunteers."

I couldn't believe Ella would knowingly volunteer for an operation that would keep her from being a mother. I knew about sterilization of mental patients in the United States, and one of the reference books I'd read mentioned that leprosy patients had been sterilized in Japan. "Did many get the operation?" I asked.

"I don't know about nobody but me," he said.

I would never know if Ella had been sterilized. I didn't want to make her sad, so I never brought up the subject again. But I felt terrible for her. She had helped me so much since I'd come to Carville. At a time when I was planning a future with my children, Ella was living out her last years. There were no children to carry on her spirit or legacy or stories. When she died, there would be no others. For Ella, it stopped here.

Just before the ten o'clock count, when I was certain Neil and Maggie would be asleep, I stood in line in the hallway and waited for a pay phone to come available. I knew Linda didn't want me to move back to Oxford. And for good reason. She wanted a fresh start and I understood. In some ways, it would have been easier for me to settle in a town without Linda, but I could not bear to be away from Neil and Maggie. As I waited and listened to an inmate yell at his wife over the phone, I felt lucky to have my children. When I was with Neil and Maggie, my troubles seemed to disappear.

If I was going to be the best father possible, I needed to live in the same town as my children. For them, and for me. But that commitment to my children came with another commitment. Since Linda would have custody, I would have to follow her wherever she moved. And I was prepared to do just that.

When the pay phone came available, I picked up the receiver and asked the operator to make a collect call to Oxford, Mississippi.

My father and me, 1972.

CHAPTER 52

I looked forward to mail call on the first of every month. That's when my father sent me $100 for my prison commissary account. In prison, $100 went a long way. I could buy decent toothpaste and mouthwash and laundry detergent, not to mention ice cream. I could also stock up on quarters so I could buy snacks from the vending machines anytime day or night, or even loan Link money for his card games. Dad had always been generous.

When I was fourteen years old, I moved in with my father. He was different from most fathers I knew. He wasn't strict. In fact, my father was the nicest dad in the neighborhood. At times, he was the drunkest.

Late at night, when drunken fathers across the country would come home to lash out at the ones they loved, my father did the opposite. He would stagger in the front door, see me in the living room, pull me up from the couch, and hug me tight. I could smell the beer on his breath and the cigarette smoke on his golf sweater. He would shake his head, as if ashamed of himself, and tell me how much he loved me, how proud he was of all I had done, and that he was sorry he wasn't a better father.

I wished Dad hadn't been so hard on himself, but I knew without question how much he loved me. I also knew he trusted me. He would not question where I had been or what I had been doing. I was fifteen. I had a car. And all the freedom a teenager could want.

Sometimes I'd sneak into my father's bedroom after my stepmother had left for work; my father would sleep until lunch. And even then, he might head to the golf course instead of his law office.

I'd pick up the pants he had thrown on the bedside chair and reach into his pocket. Most mornings I slipped in to get some spending money. I'm sure he would have given it to me if I had asked. But this was easier. And he would never notice because he wouldn't remember if he had spent it at the bar the night before.

I loved my dad, but looking at him hungover, sleeping the day away, I knew he was no model for success. He would always choose golf or drinking or hanging out with his friends over work. My idols were successful businessmen. I admired Jack Thompson, a Gulf Coast insurance man who had made so much money he sat on a board at Lloyd's of London. I looked up to my father's oldest friend, George Schloegel, who had worked his way from the mailroom of Hancock Bank to become a vice president destined for the top spot. I wanted to follow the path of Roland Weeks, the publisher of our local newspaper who spent his weekends flying stunt planes and skiing from the back of a shiny speedboat. I wasn't certain how my life would turn out, but I did know one thing. I would not end up like my father.

At mail call, two weeks after my last phone conversation with Linda, I received a large envelope. I recognized her handwriting, as well as her new Oxford return address. Inside were drawings by Neil and Maggie, a couple of photographs, and a letter from Linda sealed in a small envelope. It was the first letter I had received from her in months, and I was nervous about opening it.

I had asked a lot of Linda. I had asked her to accept my decision to live within a few minutes of her house, in her new hometown.

I would be an ex-con burdened by insurmountable debt with a $1,000-a-month restitution payment, a $600-a-month child support order, and no job.

Linda's friends, co-workers, and attorneys were giving her advice and they pretty much shared the same opinion of me. They told her to get as far away from me as possible. Once I was out of prison, they warned her, my life would be over, and I would drag her down along with me. One of her advisers went so far as to say Neil and Maggie would suffer because they carried my last name. A lawyer said I was a sociopath who would always hurt the people around me.

With trepidation I opened Linda's letter and sat on my bunk.

In a long handwritten note, Linda outlined every reason I should not move to Oxford. But in the last paragraph, her tone shifted. She would not resist my decision. Acknowledging she might be making a huge mistake, she believed Neil and Maggie needed both of us. She closed her letter with a request: *Please respect my space and privacy.*

I folded the letter and put it in my locker.

For the sake of our children, Linda was willing to sacrifice her desire to be far from me. She wanted me to be a part of Neil's and Maggie's daily lives and, by extension, part of hers, too.

I would have some hard times after my release, especially in Oxford. But Linda's blessing gave me great hope. Linda had done something remarkable. She had given me a second chance. A second chance with my children.

I didn't know how I would ever repay her.

CHAPTER 53

During the cold winter months, bundled in a brown, government-issued coat, I walked at night. I moved at a slow, methodical pace. Under the bare trees, the moon illuminated the ashy grass in the courtyard.

Ella had been right. Whatever zeal I had for uncovering the stories of the leprosy patients and inmates was inversely proportionate to my enthusiasm for discovering my own story.

So I began the process of asking myself the hard questions. *How did I get so far off course? How could I have hurt so many people? How could I have put my family at risk? Would I be able to resist the temptations of applause? Could I avoid delusional thinking and admit my shortcomings? Could I avoid caring what people thought of me? And how could I support my family in a way that did no harm, but allowed me to help others?*

I had never set aside time to look at how I felt or where I was headed. I believed I could not afford to question my motives. I was focused on a single goal—success—and had no interest in anything that stood in its way. I had convinced myself that kiting wasn't a real crime. I didn't trespass or break into buildings. I didn't speak any lies aloud or use any threats or weapons. There was no conspiracy. I'd also convinced myself there were no real victims, as long as I covered the overdraft.

But deep down, I knew better. Over the years, to ease the shame building inside, I developed a dual existence. I made a deal with myself. If I broke a rule, I would perform a good act to offset my wrong. For every instance I cheated time, I would balance the scales with an act of kindness. I was determined to do more good than bad. In Oxford, I gave free advertising space to nonprofit groups. I assigned reporters to cover news in Oxford's African American community, a group that had been ignored by the *Eagle* for a century. I wrote profiles of men and women who had committed their lives to helping the less fortunate. I gave away money and time and talent to balance the "good" side of the scale.

As the kiting became more frequent on the Mississippi coast, I again felt compelled to offer sacrifices for my deeds. I awakened early each morning to write letters of encouragement to my friends and employees. My church was without a priest, so I visited members who were hospitalized. I attended all church events. I led morning prayer when a substitute priest couldn't be found. I pledged $10,000 to the Red Cross, $5,000 to the Salvation Army, and $8,000 to the church—pledges never fulfilled.

My dual existence had worked for so long, it felt like second nature. Publicly, I was a publisher, church leader, board member, lavish boss, and budding philanthropist. Privately, I was a man who had discovered a secret technique to make money appear out of nowhere. My life became a frantic race to pile up enough good to offset my secret life.

As I walked—*meddlin'*, as Ella would say—I found no simple answers. But I did find something else. The very act of being honest with myself, taking an objective look at my life, was freeing. I felt free of the expectations I placed on myself. Free of my drive to win at all costs. Free of the prisons I'd built for myself over the past seven years.

Now I felt genuinely grateful to be here. Doc and Link checked my delusions and kept me honest; Ella and Harry were models of simplicity and kindness.

Walking among the leprosy patients—whether in the cafeteria, the hallways, or the Catholic church—helped me see my own good fortune. Watching them made self-pity impossible.

For reasons I could not fully explain, I felt an overwhelming sense of euphoria. I still did not know exactly how to change, but I had discovered some simple truths: A good life with my children did not require wealth. It was vital to be honest, without worrying about my own image. And helping others was more noble than winning awards.

I knew my elation would not last. In a few months, I would be free to make the same mistakes again. I would not have Ella to nudge me along. I would be confronted with difficult choices. And I would have to face my victims.

But now, with the sounds of crickets echoing from beneath the raised walkways, I thought about my new friends. And I was thankful my life was so rich.

CHAPTER 54

During a Wednesday night service at the Catholic church, I noticed a new leprosy patient. He sat in a pew to the right of Stan and Sarah. I'd never seen him before, but that wasn't unusual. Patients from around the world came to Carville for special surgeries and treatments. The man could have been Asian or Indian, or maybe from South America. I couldn't tell for sure, but he was performing a ritual I'd never seen. He put his Bible to his chin and pressed it against his mouth, like he was licking the pages.

Steve Read leaned over to me and whispered, "He must have skipped dinner."

When the man's face wasn't pressed against his Bible, he stared up and rocked back and forth. Then he would put his tongue back against the pages.

During Communion, standing at the altar, I got a closer look. He was blind. Like most of the victims of leprosy, the man's hands were anesthetized, so Braille was of no use. His fingertips could not feel the small bumps on the page. But he had found a new way. He was reading Braille with his tongue.

If a blind man could learn to read Braille with his tongue, surely I could find some way to make a new start. Helping leprosy patients wasn't an option for an inmate, but I could volunteer to help my fellow prisoners.

I walked to the education department and knocked on Ms. Woodsen's door. A short, rotund guard, Ms. Woodsen was also a teacher.

"May I help you?" she said, in a tone that made it clear she didn't want to be disturbed.

"I'd like to volunteer to teach a night class," I said.

"We ain't got no night classes," she said.

"Well," I said, "I was hoping I might lead a couple of classes for the other inmates."

"What you teach?"

"Speech. Debate. Public relations. Magazine publishing. Current events. Résumé writing."

"You went to college?" she asked.

"Yes, ma'am."

Ms. Woodsen stared at me for a moment. "All these men gonna need a résumé when they get out," she said. "Let me see can I get approval."

The following afternoon, Ms. Woodsen approved the night class. I was to start with résumé writing. Armed with a felt-tip pen and ruler, I stenciled a flyer advertising the class. I made copies at the prison law library, posted them at the entrance of each inmate dorm, and distributed them to inmates in the TV rooms. I even illustrated a banner advertisement at the bottom of the menu board in the inmate cafeteria. I was on a new mission.

More than thirty inmates showed up for the first class at 7:00 P.M. on a Tuesday. The classroom had twenty-five chairs, so it was truly standing-room only.

During the first hour of class, I distributed three sample résumés. We discussed the pros and cons of each. Then I reviewed the basics of résumé writing. Accentuate the positive. Use short, staccato-style prose. Organize entries in reverse chronological order. Divide accomplishments into logical categories. Proofread. Proofread. Proofread.

At the end of the hour, I opened the floor for questions.

A mortgage broker from Pennsylvania held up his hand. "How do we account for our prison time?" he asked.

"Hmmm, I'm not sure," I said. I hadn't even considered that every man in the room would have a two- to three-year gap in his professional life to explain away. "Any ideas?" I asked the class.

A New Jersey lawyer convicted of tax fraud suggested, "Just write down 'Federal Medical Center' and your job description. They'll think you worked at a hospital. Which is true."

A man convicted of insurance fraud proposed, "You could just say you worked for the Bureau of Prisons. Which is also true."

"What about this?" A counterfeiter who worked in the kitchen recommended "Chef, Gillis W. Long Hansen's Disease Center."

All the men in the room nodded and slapped hands, impressed with their own ingenuity.

"Does anyone think we should be more forthcoming?" I asked.

No one seemed to think a truthful revelation was in order.

Smitty, a Cajun marijuana grower, argued, "Hey, you said to accentuate the positive. Ain't much positive about writin' down you was in the slammer."

"I don't know," I said to the class.

"You tryin' to get us jobs," Smitty asked, "or tryin' to get us fired?"

I had envisioned this class being helpful. But not helpful in disguising our incarceration. For future classes I would propose less personal topics. Current events, debate, or maybe newspaper reporting.

Back in my dorm, I received a notice that I was scheduled for a team meeting. My team, a group of guards assigned to evaluate me, would determine whether I qualified for a furlough—five days of freedom in the midst of my prison sentence. It seemed odd that the guards would just let me go home for a week.

The next morning after breakfast, I sat in the patient cafeteria and read a copy of the regulations for inmate furloughs. Ella rolled up to my table and smiled. I told her I might get to go home for a while.

"No place like home," she said. "I goes back to Abita Springs all the time."

I was confused. I thought Ella had spent her adult life at Carville.

"When did you first get to go back?"

"I always been goin' home," she said. "'Fore my legs got bad, I walk all over town."

"I thought you contracted leprosy when you were twelve?"

"That's right," she said, nodding. "That's right."

"But you got to go home?"

"I comes and goes," she said.

There were so many things I didn't know about Ella's life. To me, her history was a puzzle with a wealth of missing parts. But at some level it didn't matter. Whatever wisdom Ella had acquired over the years was a gift she was willing to share with me. And I was grateful.

"Hope you get home soon," she said. Ella put her shiny hands on the cranks of her wheelchair and turned toward the hallways.

CHAPTER 55

I waited for my team meeting with five other inmates.

"Don't expect nothin'," one man told me. "They just fuck with you."

An hour later, my name was called. I sat in the middle of a big room surrounded by four guards. Mr. Flowers, the tall black man who dressed like a cowboy, led the meeting.

"Mr. White here had quite an offensive scheme going on the street," Mr. Flowers announced to the group.

"I see you went to Ole Miss," he said, staring at my file. "Ladies and gentlemen," he said, looking at the other team members, "Ole Miss was the last major university to integrate."

As I realized I was the only white person in the room, I remembered my first year in an all-white fraternity at Ole Miss. I was named Model Pledge. Later, as master of ceremonies, I guided new initiates through a ritual that had its roots in a secret society in fourteenth-century Bologna, Italy, when foreign students needed protection against the evil Baldassarre Cossa. Eventually, I was elected grand master.

Prospective fraternity brothers were scrutinized before receiving a coveted bid. Any shortcoming—the wrong kind of shoes, a floating eye, an unsightly mole, or any hint of a low socioeconomic background—could be grounds for rejection. When passing judgment on a potential member, a specially designed, two-compartment box was passed in silence from brother to brother. Inside one side of the box were white balls and black balls. The box allowed us to accept or reject secretly, by passing a white ball or a black ball through a small hole connecting the two compartments. One hundred men between the ages of eighteen and twenty-two sat in reverence of the ritual. To allow membership to just anyone would dilute our prestige. And that was unthinkable.

I dropped a black ball more than once to keep out an undesirable. I didn't want to hurt the young men. I just didn't want them associated with me, or with my Greek letters.

The worthy ones who did pass muster based on appearance or wealth or family reputation seemed like perfect companions. I was proud of our pedigree, and I didn't hide it. Our fraternity T-shirts read A KAPPA SIGMA: THE MOST WANTED MAN IN THE COUNTRY. The walls of the frat house were adorned with posters that read "Poverty Sucks" and "Greed Is Good."

I felt entitled and important. I was the leader of a group of handsome, affluent, prominent men who would eventually be the leaders in our state, or maybe even the country.

Now I stared at Mr. Flowers and a room full of guards who could blackball me, keep me from a furlough, and block five days on the outside with Neil and Maggie.

"If I remember correctly," Mr. Flowers said, "the National Guard had to be called in to protect Mr. Meredith."

I wanted to point out to Mr. Flowers that I was two years old during the integration of Ole Miss. I also wanted to tell him that many of the students, even in 1962, were supportive of Mr. Meredith's enrollment and integration. I wanted him to know I cochaired a Racial Reconciliation Committee and joined a group pushing to banish the Rebel flag as an official university symbol. But none of that would change his view of me. I reminded myself of Ella's advice. *What he thinks of me is none of my business.*

"Do you think you deserve a furlough?" he asked.

I was glad to see him drop the Ole Miss topic. "I don't know," I said. "But I'd love to see my children."

Then Flowers mentioned that he attended Jackson State University, a historically black institution. "We played our football games at night after the Ole Miss games," he said. "We had to sit in stands filled with the trash and beer cups left from your party."

I nodded. "We have some inconsiderate fans."

"I'm glad to see you admit this," he said.

Flowers reviewed their policy for furlough approval. "You can leave now," he said. "In the unlikely event your furlough is approved, one of us will let you know."

CHAPTER 56

I left the meeting and walked to the library. As I turned the far corner of the colony, Link joined me. We heard the rattle of chains. Guards were running. The crackle of the guards' two-way radios echoed down the hall, and Ms. Woodsen ran toward us. "Get in here!" she yelled, waving her short, fat arms. "We got an emergency census!"

Link and I waited with about thirty other inmates in the education building. Ms. Woodsen had us stand in a line against the wall as she counted and recounted us. I stood between Link and Big Gene, an inmate who weighed more than four hundred pounds. Big Gene leaned in toward me and whispered, "Somebody done left."

When a prisoner escaped, all other inmates were detained and counted to confirm the escape. To Ms. Woodsen's credit, she came up with the same number in each of her counts, but apparently a guard in another part of the colony could not get his numbers to match.

After standing for almost half an hour, Big Gene said, "Ms. Woodsen, my feets hurt."

"What'd you say?" Ms. Woodsen said, moving into teacher mode.

"I said, 'My feets hurt,'" he repeated.

"Oh," Ms. Woodsen said, "OK, I thought you said 'foots.'"

A few minutes later, Link yelled, "Ms. Woodsen, it's hot up in here."

Ms. Woodsen opened the door at the end of the hallway, and the putrid smell of chopped sugarcane rushed in. Carville, a former sugar plantation, was surrounded by sugarcane farms. When the rotten stalk is chopped into mulch, the smell can drift for miles. We all grimaced and coughed at the rancid odor.

"God," one of older inmates said, putting his hands over his mouth, "what *is* that?"

"That's Ms. Woodsen," Link said. "She opened up her ass and let some air out."

"I know who said that!" Ms. Woodsen called out from her office. "I'm gonna write you up!" At that moment, another guard yelled that the census was clear. Link ran as fast as he could back to our dorm.

As I was leaving, Ms. Woodsen emerged from her office and grabbed me by the arm. "Mr. White?" she asked.

"Yes?" I said, hoping she wasn't going to ask me to identify the inmate who said she was the source of the aroma.

"You still wanna teach?"

"Yes," I said, without thinking about it.

"Good," she said. "You start Monday morning."

A teacher. I would be helping inmates prepare for the GED test. Relieved that I would no longer have to mop floors or wash dishes or serve in the cafeteria line, I realized I would also have to relinquish my post as garnish man and menu board illustrator. I would no longer have access to the leprosy patient cafeteria. I would lose my best opportunities to talk with Harry and Jimmy. And I would lose the chance to have coffee each morning with Ella. Our time together, before the sun crested the levee, when we could talk in privacy, was coming to an end.

The patient cemetery, where many of the tombstones are engraved with aliases.

CHAPTER 57

Link was given a new job, too. The guards, who knew he was scared of ghosts, gave him a job weed-eating the grass around the tombstones in the leprosy graveyard.

The cemetery, covered in shade from pecan trees, was located at the back of the colony. Leprosy patients, most of whom had been abandoned by family, were buried there. With more than seven hundred white tombstones similar to military markers, the graveyard was visible from the second-floor hallway. As dawn came, the tips of the stones jutted out just above a shallow mist. The tombstones were often marked with nothing more than a patient number. When a name was engraved, it was often the patient's alias etched in marble.

After his first day on the job, Link said he stepped in a sinkhole and his leg sank deep into the ground.

"I felt the goddamn leper bones!" he yelled. I didn't believe him, but Link was superstitious. After the "leper bones" incident, he refused to set foot inside the graveyard. Later that week, a guard asked Link to climb under one of the buildings to repair a pipe. Link had heard about an inmate who, climbing under the same building, had been bitten by a snake. On the neck.

"But it wasn't poisonous," I said.

"Goddamn!" Link said. "It don't matter. A snake bite me in the neck, I'm gonna have a motherfuckin' heart attack!"

The following morning, two guards tried to coerce Link into going under the building. One of the guards, in a fit of frustration, tried to push Link's head under, and Link bit his hand. Free of the guard's grip, he raced around the colony screaming, throwing his hands in the air, trying to bite anyone who touched him. When they finally caught him, he was sent immediately to the hole.

As the Bureau of Prisons geared up to take over the colony, the inmate population surged. A new crop of inmates would be transferred to Carville from other prisons, and they would take most of our jobs in food service. Doc got his hands on an internal memo that listed Carville as 99 percent over capacity. In order to abide by federal regulations, the warden would have to take over the entire facility. And that's what he wanted.

I told Ella about my new job. She nodded, as though she approved. I wanted to know if she had heard about the future of her home. I wondered if the patients had been informed about the Bureau of Prisons's plans. I couldn't imagine the stress relocation would bring for long-term residents like Ella who had spent their lives here, where they were safe.

If Ella were relocated, I might not see her again. I found myself scheming about ways to pass her notes in the hallway. If I mailed her letters, no one would ever know. I could send letters through the prison mail system, and a few days later, it would arrive at Ella's post office box. Even if she were relocated, we could keep up a correspondence.

For decades, mail at Carville was controversial. Until the late 1960s, all mail from the colony was sterilized before being released into the general population. It was baked in a huge electric oven. Even *The Star*, the magazine published by the patients, was cooked after it rolled off the press to be mailed to forty-eight states and thirty countries.

One issue of the magazine, almost eight thousand copies, was burned to a crisp because a staff member was inattentive. I would have been outraged had my magazine been charred before subscribers received it.

"Do you write letters?" I asked Ella.

"Nope."

Many of the leprosy patients didn't write letters. Holding a pen or pencil was difficult; for those with severe absorption, typewriters weren't much help, either.

"Do you get any letters?" I asked.

"I gets some," she said, "but I don't look at 'em." Ella said she gave them to a lady friend who read the important ones. I didn't want anyone else reading my letters to Ella. And, if the "lady friend" was an employee of Public Health Services, as I suspected she might be, she might report me to the guards.

Jimmy Harris sat down at our table. "Good morning to you both," he said, nodding to me and Ella. Jimmy was the one patient who would talk about any subject. I felt like nothing was off-limits with him.

"Jimmy," I asked, "where was the oven—where they baked the mail?"

"Across the way," he said, pointing toward the infirmary. "Every piece got the treatment. Sometimes they burned it."

Jefferson, the skinny inmate with gold teeth from New Orleans, danced by our table. "They baked your mail," he said, "but that ain't shit. You shoulda seen what they did at the Loyola Street Post Office in New Orleans."

"What happened there?" Jimmy said.

"They zap them letters with an X-ray machine."

Jimmy and Ella were confused. "What you talkin' 'bout, boy?" Ella said.

Jefferson told us he'd worked for the postal service in New Orleans.

"You delivered mail?" I asked.

"Naw, I worked at the main post office building. I manned that X-ray machine." According to Jefferson, every single letter and package went through the X-ray machine before being processed.

"Ever see anything interesting?" I asked.

"Every damn day," Jefferson said. "When I see cash money go through, I just pick up the letter and stick it in my back pocket."

"You can see cash through an X-ray?" I asked.

"Hell, yeah," he said. "It stands out like *you do* in prison."

Ella laughed. Jimmy smiled, too.

"But people don't send cash that often, do they?" I asked.

"Shit," Jefferson said, bobbing his head up and down and laughing, "I loved them holidays. Christmas and Easter were the best. People be sending money to everybody they know. I made $900 one day before Christmas."

Jimmy looked appalled, and Ella shook her head.

"You didn't feel bad about taking people's Christmas money?" I asked.

"Hey, I got a heart," he said. "I took the time to read every one of them cards. Like on Mother's Day, nicest notes you've ever seen. One said: *Mom, your family loves you. I hope this comes in handy.* And it sure did!" Jefferson laughed.

Ella scolded him for taking money from mothers. Jefferson acted as if he didn't hear her. The more he talked, the more excited he became. "Birthdays was better," he said. "I loved birthdays. I'd take them cards home and read every one of 'em. Then I'd count the money and have me a birthday party!" Jefferson started singing: *Happy birthday to me. Happy birthday to me. Happy birthday, dear Jefferson. Happy birthday to me!*

As Jefferson sang, I thought about the birthday cards I'd sent to Neil and Maggie and how I would have felt if someone had taken them. "Tell me you didn't take money from little kids."

"Sure did!" he said, proudly. "One day I got this card, it say: *Tommy, here's a brand-new $100 bill for every year you've been alive.*" Jefferson smiled. "And that kid was five years old."

"That's terrible!" Ella said.

"It sure was," Jefferson said. "I was hopin' he was a teenager." He laughed and danced some more. Jefferson said he had been named employee of the month on several occasions.

"I was always happy to work overtime on the X-ray machine," Jefferson said. "They just loved me down at the Loyola Street Post Office."

"I bet they didn't love you when you got caught," I said.

"Got caught!" Jefferson said. "I'm in here for money orders. Those dumb asses never knew about the X-ray machine." Jefferson told us he had saved more than $40,000 from working the X-ray machine. "I'm goin' legit," he said. "I'm buyin' me a franchise when I get out."

He turned and danced into the kitchen. Ella shook her head.

"Mail wasn't the only thing they baked in the old days," Jimmy said. "They tried to cook me!"

Jimmy was one of a handful of patients who in the 1930s volunteered for an experimental fever treatment. Jimmy and the others were transported to an isolated wing of a marine hospital in New Orleans. There, they were placed in fever machines. Ella and I listened to Jimmy talk about the nurses, about spending Christmas in the hospital, about the doctor who pushed boundaries to raise his body temperature higher, and the fever machine that looked like an iron lung.

"The contraption had two openings," Jimmy said. "One for my head. The other for the thermometer in my rectum." At that, Ella turned away and rolled toward the coffeepot.

"Did it work?" I asked.

"Just might have," Jimmy said. "But it almost killed me." During his final treatment, Jimmy's body temperature exceeded 105 degrees. He blacked out. The nurses told him later he had gone berserk. They stopped the treatments and sent everyone back to Carville. But a year later, Jimmy was released. Jimmy had tested negative for leprosy twelve months in a row.

Jimmy's story corroborated Doc's theories of thermogenic treatment. The heat had nearly killed Jimmy, but it worked. Doc's heat pill might have been dangerous, but it seemed like a pretty good alternative for those with terminal illness. And, not unlike chemotherapy, many medical procedures almost kill the patient in order to kill the foreign object.

"If you were cured," I asked Jimmy, "why did you come back?"

"Leprosy came back," he said. "Stubborn little bug."

Jimmy moved back and forth between Carville and home. But it was his wife who needed treatment most of the time. Jimmy built a life and career away from the colony. During his first stint at Carville, a nun named Sister Hilary encouraged him to use a camera. So when Jimmy tested negative for leprosy, he made a living in Ville Platte as a professional photographer.

"I did more than seven hundred weddings," he said.

"Did the people in Ville Platte know you'd been in Carville?"

"Most did," he said.

"Did that affect your business?"

"Never!" Jimmy said. "And when I started shooting in color, I went to the bank every day. Not to borrow, mind you, but to put money in."

Jimmy and his wife still came to Carville to get treatments, but they were outpatients, having lived independently for almost three decades.

"So why are you back?" I imagined his leprosy had returned.

Jimmy said his wife's feet, devoid of feeling, were in terrible shape. She required daily treatments.

Jimmy leaned in toward me. "Most people won't tell you this," he said in a low voice, "but any of us who were forced to be here can come back to live." He looked around to make sure no one could hear what he was about to say. "It doesn't cost us a cent." He sat back in the chair. "Not a bad deal, if you can get it."

A guard came into the cafeteria. "Where's Jefferson?" he asked me. I said I had no idea. "Go find him," the guard said.

I found Jefferson asleep in the back of the cooler. We walked together back to the leprosy patient cafeteria.

"You got a new job assignment," the guard told Jefferson.

Jefferson started dancing and singing. "I ain't never washin' another dish as long as I live!" The guard escorted Jefferson toward the hallway.

"Hey, Jefferson," I called out, "where's your new job?"

Jefferson looked over his shoulder and smiled, "Prison mail room!"

CHAPTER 58

"Hey, Harry," I said, "this is my last day."

"You goin' home?"

"No, changing jobs."

Harry tipped his hat, like he always did. "Good luck," he said.

"Do you have a job?" I asked.

"I get minimum wage," he said.

"What do you do?"

"I help out," he said. Harry assisted other patients. He pushed their wheelchairs to and from the infirmary. He rescued patients with a dead wheelchair battery who might be stranded around the colony. He ran errands. "I do what they ask me to do," he said.

No fanfare. No complaints. The work was simple. It was quiet. In fact, I hadn't even noticed his tasks were jobs. It looked more like a routine. He helped people. Each little rescue was important. Harry spent his days doing small, great acts.

I would miss seeing him in the cafeteria, watching him eat while using his special utensils. He was as friendly as any person I had ever met. But I was curious about one thing. He never called the inmates by name. He would tip his hat and give us a smile, but he never did use our names, or even nicknames.

"My name is Neil," I reminded him.

"I know," he said, like I had insulted him.

When I asked why he never called any of the prisoners by name, he looked down at the floor and shuffled his Velcro shoes. Embarrassed, he said, "You all sort of look alike."

When Ella arrived for breakfast, I poured her a cup of coffee.

I felt a love for her like I'd not felt before. Not quite like I loved my siblings or my wife or my parents. I respected her and she treated me with respect, even though I was a convicted felon.

"Saturday will be my last day," I reminded her. "I'll miss having coffee with you."

"Miss you, too," she said, smiling.

I would see Ella in the hallways, in church, on our daily walking routes, but I would miss our early mornings together.

"I'll miss our talks most of all," I said.

"We'll talk," she said.

"Where?"

"In the breeze," she said, staring off into the leprosy patient courtyard.

I nodded like I knew what she meant, put my hand on her shoulder, and told her I'd see her later. As I walked toward the kitchen to turn in my apron and dry erase markers, Ella stared out the window. "Yep," she said in an airy voice, "I'll see you in the breeze."

CHAPTER 59

On the morning before my first day of work as a teacher, about fifteen men lined up outside the door of my prison room. I'd never been in the room at 8:00 A.M. on a Monday morning. I'd always been mopping the floor or writing on the menu board. One at a time, the men walked into our room, stood in front of Doc's bunk, and described their symptoms. Doc would listen, look down their throats or feel underneath their jaw, and jot down a few notes. Then, he would tell them exactly what to tell the physician assistants they were to see later that morning.

"Clark Kent," one of the inmates said, "you didn't know Doc here saved my life last month."

Doc had caught a mistake made by the prison doctors. A deadly combination of drugs had been prescribed, inadvertently, by two different physicians.

The men who came into our room to be examined were young and old, Christian and Muslim, black, white, and Hispanic. For all Doc's talk of not wanting to be around these men, he still honored his Hippocratic oath. He examined the men, made a diagnosis, and sent them on their way. Not a penny exchanged hands. Doc was full of surprises.

When I arrived in the education department, Ms. Woodsen sat at her desk in the corner of the room. I had convinced her to let me take the lead in helping the inmate students pass the GED. The Bureau of Prisons received money for each inmate who graduated, and I was certain I could teach them enough to pass a high school equivalency test. Never one to set the bar low, I had a secret goal of 100 percent graduation, but I told Ms. Woodsen I thought I could achieve a 50 percent graduation rate.

My strategy for success was simple. I would start with questions. I would discover what the men did *not* know. That was the key. I introduced myself to the class and told them about my background. I emphasized that questions and curiosity were the secrets to learning. I wanted them to be comfortable asking me any question. I waited for someone to speak up, but they were slow to ask.

The classroom was filled. Several students were my friends. Ricky, a handball buddy, sat on the front row. Mr. Dingham, a union boss from Newark who had signed labor deals for thousands of stevedores at the port, sat in the back. I was surprised he didn't own a diploma.

"C'mon," I said, smiling, trying to make them feel at ease, "you must have questions about something."

Mr. Dingham raised his hand and said he had two questions. "Is it true," he asked in a strong New Jersey accent, "that vultures ain't got no assholes?"

Ms. Woodsen screamed at him from across the room. "You can't use them words, Mr. Dingham!"

Dingham yelled back. "I heard they just puked up everything and don't never need to take a shit."

Ms. Woodsen stood up and pointed a chubby index finger at him. "We use scientific language up in here!"

I stepped between them and assured Mr. Dingham that what he'd heard about vultures was a myth, but that it might be true that vultures regurgitated. Even scavengers had trouble digesting bones and feathers, I explained. Then I demonstrated to the class how one would rephrase Mr. Dingham's question, scientifically.

"Is it true that vultures don't have a rectum?" I recited. "Because I heard they regurgitate their wastes and, consequently, do not excrete feces."

Ricky, my handball buddy, was confused. "Feces?" he asked.

"You know," I said, pointing to my own rear, "poop. . . . dook." Then tentatively, "Shit."

Ricky smiled like he understood. Then he leaned over to his friend. "You're a feces head."

I jumped in and told Mr. Dingham to ask his second question—reminding him to avoid curse words.

"Well, I've always wanted to ask . . . ," Mr. Dingham hesitated.

I assured him it was fine. "There's no such thing as a bad question."

"Well . . ." he said, "I've always wondered . . . what's a vowel?"

The other men in the room didn't laugh or snort. And no one rattled off the letters. They were waiting for my answer, too. I turned my back to the class and wrote *A, E, I, O, U, and sometimes Y* on the chalkboard. To reach my goal, I would have to spend hours tutoring after class.

This was going to take longer than I'd thought.

CHAPTER 60

As the gray winter months lingered, the leprosy patients became more and more anxious about their future. Word spread about an inmate who had filed a federal lawsuit against the Bureau of Prisons for exposing him to leprosy patients. A preliminary trial date had already been set in Baton Rouge, and the documents filed by the inmate's attorney included inflammatory language about unsuitable, dangerous conditions of a minimum-security facility where federal convicts and leprosy patients were free to mingle.

We all knew this would bring the battle for control of Carville to a head. The Bureau of Prisons, intent on taking over the entire colony, had offered to help pay to relocate the patients. The patients weren't sure if they would be offered some form of compensation, or if they would be provided a new building somewhere else on the three-hundred-acre colony, or if they would be transferred to a nursing home. Their fate was uncertain, and the Bureau was not sharing the details. Or asking their opinions.

The patients were suspicious from experience. In the early days, they were kept in the dark about decisions on furloughs, new treatments, and patients' rights. Privileges promised by one director would later be revoked by his successor. And recently, the patients had been misled by the Bureau of Prisons. They were told that only geriatric, invalid prisoners would be housed at Carville. They had no inkling that more than 250 healthy, possibly dangerous prisoners would be a part of the arrangement.

But the real problem was money. The $17 million budget to maintain 130 patients at Carville was scheduled to be cut.

Money—and currency—had always been a problem for leprosy patients. Most colonies around the world minted their own coins and currency to prevent leprosy patients from amassing money to fund escapes, as well as to avoid infecting the general population. The coins usually honored Lazarus or the current leader of the country. But Carville was never required to produce its own currency. In the early days, cash was fumigated with disinfectants. Later, checks and cash mailed by patients were baked before circulating among the general public. Within the colony, though, money caused problems. A staff person who performed tasks like radio repairs for a patient might be wary of taking contaminated currency. On more than one occasion, patients who paid for repairs would later discover that the bills had been run through a commercial washing machine and hung out to dry on a clothesline. Even the people who worked at Carville were afraid. And the ones who weren't wanted to keep up pretenses.

The staff at Carville was paid a 25 to 50 percent premium for working at the colony. It was called hazard pay—a premium for risking their well-being to work at the hospital. Stanley Stein and the staff at *The Star* had fought to eliminate hazard pay. This battle pitted the Carville crusader against the very people who cared for him. Stein became quite unpopular among some of the staff. In the end, a nomenclature compromise was reached. Pay would not be cut, but it would be renamed something more innocuous.

On a Sunday morning after a rare winter frost, Sarah and Stan talked to Father Reynolds about their growing anxiety. Sarah was worried about being placed in a strange nursing home.

"Can you imagine their reaction?" she said. For blind patients, any relocation would mean learning to navigate new paths.

Ella was worried, too. After church on a Sunday afternoon, she was waiting for me at the entrance separating the prison side from the leprosy side. I made sure no guards were around and stepped into the screened corridor. It was the first chance we'd had to talk since I had been transferred to the education department. Everyone else was inside watching football playoffs.

"Hey, Ella," I said, "what are you doing here?"

"Settin' in the breeze," she said.

And she was. There was no wind, but she was sitting in the spot the patients called the breezeway. This was where we would talk. Ella asked about my new job, and I asked her about the new menu board guy. Then Harry rode up on his bike. He stopped and shook his head. He couldn't believe he was going to be relocated. Harry had lived at the colony since 1954. Other than a furlough each year to see his mother in the Caribbean, Harry had spent forty years at Carville.

I wished I could help put a stop to the plans. I tried to be encouraging, pointing out the most obvious benefit, the proposed $33,000 annual stipend.

"If the stipend comes through," I said, trying to be upbeat, "you could move into a house. In a neighborhood."

"This here is your prison," Ella said, "but it's our home."

Harry shook his head again, and for the first time since I had known him, he frowned. He stared at his shoes. "People back away," he said. He grabbed the brim of his hat with his two good fingers and placed his mitten hand against the back. He pulled it down over his brow and muttered, "Never get used to it."

"People don't want folks like us stayin' on they street," Ella said.

"That's not true," I told them. "People would understand, once they got to know you." I pointed out that we were neighbors, right now. "I'm honored to live next to you. And I would have been on the outside."

Ella looked at me, skeptical. *"Mmm uhh!"*

Then I remembered Lionel Day. In 1973, Lionel moved down the street from my house. He was the first African American kid to live in our neighborhood. His father owned gas stations. They bought a two-story house in all-white Bayou View, two doors down from where my grandparents once lived. I overheard adults talk about plummeting property values. I heard one mother say, "The nerve of those people."

A week after his family moved in, Lionel left for school and saw a FOR SALE sign in his yard. I don't know if kids did it as a prank or if adults did it for more ominous reasons. But the message was clear.

I listened as some of the older boys in Bayou View plotted more pranks—deflating the tires of their car, lining their driveway with watermelons, spray-painting their lawn with the word *nigger* in green, glow-in-the-dark paint so it would show up only at night. And when they laughed at their own cleverness, I pretended to laugh right along with them.

Lionel was my friend. We were both on the student council. I wanted to make him feel welcome in our neighborhood. I wanted to knock on his front door and invite him to my house. I wanted to apologize for the actions of my fellow whites. But I didn't do any of those things.

I was afraid of the older kids. Afraid of the names I'd be called. Afraid to be on the outside.

"I'm not leaving," Harry said, serious and assertive. Sweet Harry, with the great straw hat, all of a sudden seemed sturdy, forceful.

Ella nodded. "I ain't leavin' neither."

I tried to imagine Ella and Harry living on the outside and how the neighbors might react to their missing legs and absorbed fingers. But they would carry much more into the neighborhood than their disfigurement. Ella and Harry would be found out. People would discover a "leper" had moved into the neighborhood. A FOR SALE sign in the yard might be just the beginning.

I had done nothing for Lionel Day. I might have been honored to be his friend, but not enough to stand by him publicly.

I understood perfectly why Ella and Harry refused to leave Carville. The world, out there, was full of people like me.

CHAPTER 61

As the Bureau of Prisons continued preparations to take over the colony and evict the patients, dozens of new, sick inmates were transferred to Carville, men with amputations, spinal injuries, and failing organs. Among them were a young boy, not much past eighteen, who weighed over five hundred pounds, and a man with an enormous right leg that was nearly three times the size of his left. It extended straight out in front of his wheelchair.

I stood in the hallway with Doc and Steve Read. We watched as he was pushed down the hallway by an orderly.

"Looks like elephantiasis," Doc whispered.

"That's a form of leprosy, isn't it?" I said.

"Hey," Steve said, "he's got dual citizenship."

The prison population swelled, and rumors about the fate of the leprosy patients continued to swirl. The lives of my Carville friends, on both sides, seemed to be unraveling, just as I was beginning to feel hopeful about my own.

Fights were rampant on the inmate side. Sergio, the Cuban who had baked sweets for Neil and Maggie, was hit in the face more than a dozen times because he switched channels in one of the TV rooms. His lacerations required forty stitches, and he was thrown in the hole for fighting.

A skinny black kid named Calvin, who had been a Golden Gloves boxer as a teenager, pummeled one of the obese inmates to the point of tears. The big guy had snitched on Calvin for selling stolen batteries.

Two inmates in their seventies attacked each other with five-pound dumbbells. On the same day, a small disabled inmate used a walking cane to open a huge gash in the back of his roommate's head.

The hole, the real jail inside Carville, was full. Anyone who ran into serious trouble now would be sent, temporarily, to the Iberville Parish Jail.

CeeCee was caught having sex with her new boyfriend in one of the janitorial closets. She was shipped off to the parish jail, along with her new friend.

Link, who had already been in the hole for biting, continued his quest to avoid work. If guards dragged him to the landscape department, he found a comfortable spot at the base of a tree and took a nap. If the guards looked away, he slipped back to his room and climbed under his blanket to sleep. Finally, the guards shipped him off to serve two weeks in solitary confinement at the parish jail.

Dan Duchaine was caught recording bodybuilding audiotapes over the pay telephones. His telephone privileges were taken away. If a guard caught him using the phone again, he, too, would be shipped off to join CeeCee and Link.

Doc was also having trouble. He lost his visiting privileges. His girlfriend had tried to smuggle Super Glue into the visiting room so Doc could perform some dental repair work. The guards told him his visitation rights might never be reinstated.

Even Steve Read was having difficulties. He started finding small piles of his hair on his pillow. He blamed the water. Carville was in the heart of Cancer Alley, a thirty-mile section of the Mississippi River where petrochemical plants dumped their toxins in the ground. "It's cancer," he said, "I just know it." He also had a strange rash on his torso; he was sure he had leprosy.

In addition to the fights and loss of privileges and confinement in the hole, tragedy struck both sides of the colony. On the inmate side, José, a jovial Puerto Rican who had been imprisoned on corruption charges, died in his sleep. A bank president from Florida died of a massive brain aneurysm. And an inmate from Texas, after hearing of his wife's affair, tried to commit suicide by drinking battery acid. A helicopter from a Baton Rouge hospital transported him out late at night.

Things weren't much better on the leprosy side. A friendly, often drunk, leprosy patient disappeared from the hallways. When we asked a staff member about her whereabouts, he said, "She passed." One morning on the way to the cafeteria, I noticed that another leprosy patient had recently lost a leg.

About the only break from the tension was the Carville Book Club. The prison librarian had helped two graduate students from the English Department at Louisiana State University finagle permission to lead the discussions. The first book we read was J. D. Salinger's *Catcher in the Rye*. I arrived at the classroom early. The two grad students were sitting at the front of the room. The woman, Nancy, had rather large breasts and wore a shirt two sizes too small. The man, slender and effeminate, introduced himself as "Tater." While the three of us waited for the other inmates to arrive, they asked me a few questions about my life before Carville, but they didn't mention books or literature. I wondered whether they had come to talk books or to meet men.

The first discussion didn't go so well. Dan Duchaine was in a foul mood. He kept referring to Salinger as a "latent fag." I felt bad for Tater and tried to steer the discussion in another direction, but Duchaine was relentless. After about twenty-five minutes, when the discussion waned, Nancy polled the group. More than half the men admitted they hadn't bothered to read the book.

"You'd think we have plenty of time on our hands," I said, trying to smooth things over. Nancy and Tater smiled. I was beginning to feel like the teachers' pet.

We met for four weeks in a row. We read a collection of short stories, a southern novel, and *A Lesson Before Dying*, by Ernest Gaines, a Louisiana native. The book revolved around the execution of an innocent but illiterate black man in Louisiana. Capital punishment seemed to resonate with the group. Then, at the end of a robust discussion, Nancy and Tater said they wanted to try something new. They asked us to answer, aloud, three questions: *What was your crime? What is your favorite book? What is your sexual fantasy?*

I knew then Nancy and Tater weren't here for altruistic motives, and I was pretty sure it would be their last week leading our book club. I noticed Mr. Povenmire, the guard who was in charge of the education department. He stood just outside the door listening to the discussion.

When my turn came around, I hesitated. Then Duchaine jumped in: "As English majors, you're gonna love this guy's crime."

"What'd he do?" Nancy asked.

"He's a creative check writer!"

CHAPTER 62

Back from two weeks in parish jail, Link had found a new object of his affection, a young, quite beautiful leprosy patient from Brazil who had just arrived to seek treatment at Carville. She was tall and blond and curvy. Her clothes were tight. When she walked down the corridor between the inmate side and the leprosy side, she got lots of attention.

Link yelled, "Will you marry me!?"

The woman smiled and waved at Link. She rounded the corridors several times a day.

"I'm in love!" he screamed.

After the woman was out of sight, Link walked over to the bench where Frank Ragano and I were reviewing a dummy cover for his forthcoming book, *Mob Lawyer*.

"That bitch fine," Link said.

"Good God, man!" Frank said. "She's got leprosy!"

"I'd fuck her," Link said. "She only got half a foot, but I'd still fuck her!"

Link ran off to wait for her next pass.

Frank handed me a letter he had received from his publisher, Scribner. "What do you make of this?" he asked. Frank had written a book about representing notorious criminals, but the publisher wasn't completely happy with the manuscript. Frank knew I had been a magazine publisher and, occasionally, he would ask my opinion.

The letter explained that the editors had decided to alter the format of his book, to publish two parallel narratives. One written by

Frank; the other written by Selwyn Raab, a reporter of such distinction that his writings had inspired the television series *Kojak*. Scribner intended to list Frank Ragano *and* Selwyn Raab as authors. Frank had confided to me how much he relished seeing his photograph in the newspaper or on the evening news. He didn't want to share credit for his book.

"They think enough of your writing skills to keep the manuscript intact," I said. I explained that, albeit unusual, it was better than having a book written *with* another writer. What I didn't mention was that Raab's alternate voice was certainly the publisher's way of saying *we question your credibility*.

Frank had also received a proof of the book's cover art. The cover was black with white typography. The title, *Mob Lawyer*, was in bold white script with red ink seeping into the letters, like blood might soak into a white shirt. The subtitle of the book read *Including the Inside Account of Who Killed Jimmy Hoffa and JFK*.

"A good color combination," I assured Frank.

Frank was waiting for his wife and son, who were coming to visit. While he waited for them to arrive, I asked him about the JFK assassination.

"It started with a message," he said. Frank believed he unwittingly delivered the message from Jimmy Hoffa to Carlos Marcello to kill Kennedy. Later, his belief was confirmed by a deathbed confession of a Tampa Mafia boss.

"I wish I had never met those people," he said. Born to a struggling merchant in the poor section of Tampa, Frank could never get enough. And that led to terrible regret. Regret that he had helped his clients continue to do horrific acts. Regret that he had become known as the "mob lawyer." Regret that he had become a target of prosecutors and was spending one of his last years in a place like Carville.

Frank and I sat quietly for a moment; then he told me another story.

The CIA had partnered with one of his Mafia clients. In an attempt to bring Castro's reign to an end, the CIA looked to Mafia bosses who stood to lose money if Castro were to stay in power. According to Frank, the CIA gave Santo Trafficante hundreds of thousands of dollars, along with poison pills, to kill Castro. Santo took the money and flushed the poison pills down the toilet. He used the payoff for his business interests in Cuba and told the CIA the assassination attempt had failed.

Then Frank mentioned that the Public Broadcasting System had just released a one-hour documentary about his life as lawyer to the Mafia.

A week later, Frank and the prison librarian received permission from the warden to show the documentary to the inmates. Filmed by PBS's *Frontline*, the documentary was entitled "JFK, Hoffa and the Mob." We made arrangements with the guards to show the one-hour film in the large classroom as a part of the current events class. I made flyers to promote the occasion and persuaded the new menu board guy to plug the event in the cafeteria.

About thirty inmates attended, including Steve Read, Doc, Art Levin, Dan Duchaine, and Frank's best inmate friend, Danny Coates. A couple of dozen counterfeiters, tax evaders, swindlers, and drug traffickers also attended, as did a new arrival at Carville—a computer whiz named Gary, who, at age twenty-four, had tapped into the Federal Reserve system and wired himself $125,000. As I looked around the room, I thought there probably should have been a law against this industrious group's convening, but no guards were in sight.

Art Levin, the man who had watched over Carlos Marcello and helped me beat Steve Read in Monopoly, sat in the back of the room as Frank introduced the video.

The film featured Frank Ragano as the intimate friend and lawyer to Teamster president Jimmy Hoffa, as well as attorney to Santo Trafficante, one of the most feared Mafia bosses. The documentary asserted that Ragano was the first mob lawyer to go public with what he knew. During the interviews, Ragano recounted mob involvement in CIA plots to kill Castro. He alleged that the Mafia had orchestrated the murder of Jimmy Hoffa and the assassination of John Kennedy, and he admitted, on videotape, to "toasting" the death of JFK. In the end, Frank told the interviewer he had unwittingly delivered the message from Hoffa via Trafficante to Marcello to have JFK killed.

It was a sobering moment. There was silence in the room as the credits ran.

I turned on the lights. Frank asked if there were any questions.

Doc raised his hand. "How much did you charge these guys?"

Frank said he charged Jimmy Hoffa about $40,000. "I never charged Trafficante anything." Doc looked suspicious, as if he couldn't imagine performing a professional service gratis.

Steve Read blurted out, "I have a two-part question, Frank: one, was JFK the first president you knocked off? And, two, do you have your sights on Clinton?"

One of Frank's friends yelled back, "What about you, Read!? What country music star are you gonna kill next, Dolly Parton?"

The Q & A session broke into a series of sarcastic exchanges that led to a yelling match. I reassessed whether these guys were such a formidable bunch after all.

Then I noticed Mr. Levin. In the midst of the arguing and insults, he sat still. Mr. Levin was counsel to Carlos Marcello. Marcello allegedly handled the details of the assassination. Mr. Levin helped Marcello navigate the Louisiana legal system and operate within the bounds of the law. As one of Carlos Marcello's closest confidants, he was privy to the details of Marcello's business.

As the other inmates argued, I thought about all Mr. Levin must have known. Frank Ragano might have been the first mob lawyer in the country to go public, but the one who probably knew the most sat quietly in the back of the room.

CHAPTER 63

I returned to my room to find Doc burning a growth off his torso.
This time he had inserted the tip of a sewing needle into a large nod-
ule on his stomach. He held a burning match under the needle to heat
it. Doc noticed me staring.

"Needle's not a bad conductor," Doc said, as he lit another match.
I got the first whiff of burning flesh when four guards marched into
the room.

"Goddamn!" one of the guards said when he saw Doc's procedure.
"What the fuck you doin', Dombrowsky?"

Doc lit another match. The guards hadn't come for Doc. They had
come for me.

"White," a female guard said, "go stand in the hall."

"Why?"

"Get out!" she screamed.

As I stood in the hallway, the four guards emptied the contents of
my locker and spread them out on my bunk. The procedure was called
a "shakedown." They examined my belongings. They uncoupled my
socks, shook each article of clothing, tasted my toothpaste, smelled my
bottles of shampoo and conditioner, and flipped through the pages of
my books, magazines, and diary notes. They found my stash of scent
strips from the magazines and confiscated them as contraband.

"Mr. White," the female guard called out.

I stepped into the doorway. "Yes," I said. My heart rate shot up.

"You hoardin'!"

"What?"

"You got ten sheets of carbon paper!" she said. "That's hoardin'."

Doc, who had been allowed to stay in the room, blew out his match and interjected, "He really likes office supplies."

I wrote four to five letters a day, and I made a carbon copy of any I might not retrieve after my release.

"Why can't I have ten sheets?" I asked.

"You can tattoo yourself!" The female guard explained that I could press a thumbtack through the carbon paper and into my skin to duplicate the effects of a tattoo needle.

I pulled up my sleeves to show her my clean arms, and she laughed. The thought of me tattooing myself seemed ridiculous, even to her. Doc lit another match.

"I keep duplicates of my letters," I said.

The guard said that if I told her I would use all of the carbon paper within forty-eight hours, I could keep it.

"It should last longer than that," I said.

"But if you *tell me* you will use it in forty-eight hours," she said, "I can let you keep it."

One of the other guards, trying to be helpful, said under his breath, "Just lie."

"I'm in here for lying," I said.

The guards looked at each other for a moment and went back to the shakedown. I stepped back into the hallway and leaned against the wall. For so many years I had used my connections to get special treatment. I expected people to overlook my tendency to bend the rules or cut corners or even kite checks. Standing in the hallway, temporarily banished from my prison room for illicit possession of office supplies, I felt good about telling the truth. It was a small thing. Nothing at risk but a few sheets of carbon paper. But it felt important.

CHAPTER 64

On a Sunday afternoon in late January, more than forty-five inmates packed into the sports TV room to watch a playoff game. All the seats were taken, so several men, including Mr. Dingham and his friend John Gray, leaned against a windowsill. Dingham had a terrible cough. He hacked and coughed, and then hacked and coughed some more. After about fifteen minutes, Juan, the wheelchair-bound Mexican inmate who had been shot by a DEA agent, told Mr. Dingham to leave the room if he couldn't control his cough.

"I can't help it," Dingham answered. "I'm sick."

"Then shut the fuck up!" Juan said. He and Dingham knew each other well. They were classmates and my students.

The coughing continued, interrupting the commentary and irritating everyone in the room. Juan, who had been cantankerous all week in our classroom, couldn't take it any longer. He picked up his thermos mug and tossed coffee at Mr. Dingham, except it landed on the man next to him, John Gray.

John Gray stood and walked toward Juan.

Fights in television rooms had come to be expected. Any inmate who spent any time watching television would eventually witness a fight. John Gray didn't want to fight a man in a wheelchair. He expected an apology.

John Gray stood before Juan, held out his coffee-stained T-shirt and said, "What the hell did you do that for?!"

John Gray did not get what he expected. Juan, still in his wheelchair, thrust his right hand at John Gray's chest. It wasn't really a punch or a push or even a slap. Later, I heard it described as a jab.

Juan turned his wheelchair around and zipped out of the room. John Gray put his hands over his sternum and fell to his knees. Kirk, an inmate from Lafayette, Louisiana, who wrestled alligators, tried to help him up.

"He got me," John Gray told Kirk. Blood had started to soak through John Gray's T-shirt. When Kirk saw it, he bolted from the room and ran down the hallway. An inmate in the TV room yelled, "Shank!"

Juan rolled down the hallway as fast as he could, expecting to escape from prison by wheelchair. Kirk tackled Juan from behind, knocking him to the floor. Juan flopped like a fish on the concrete, waving his homemade knife, willing to cut anyone who got near him. Kirk eventually wrestled the knife away from Juan, and in the process saved a guard who had just stepped through a doorway right over Juan.

Kirk controlled the situation. He tossed the shank out of reach and pinned the paraplegic to the floor until the guards handcuffed him.

The shank, a sharpened piece of metal, lay on the concrete floor. Drops of blood spotted the hallway. The tip of the knife was covered in the brightest red blood I had ever seen.

Juan had stabbed John Gray over the smallest act, a twist of fate. I thought about how many times I had leaned over Juan to help him with his math, and I wondered if his shank had always been within reach. I imagined how many times Ella and Harry had passed within one foot of Juan in the hallways.

As the guards dragged Juan and Kirk to the hole, I stared at John's blood against the dull colors of the colony floors.

CHAPTER 65

The prison alarm echoed through the hallways, and the guards rushed us all to our rooms. When the siren was finally turned off, I heard the sound of the ambulance helicopter that would take John Gray to a hospital in Baton Rouge.

Overnight, Carville became a high-security prison. Our freedom of movement was taken away. We were confined to our rooms. Extra guards were hired. They wore black T-shirts with the letters SWAT printed on the back.

The FBI was called in to investigate. They interviewed everyone who had been in the room. The Mexican inmates said they hadn't seen anything. A few of the white witnesses told the agents exactly what happened. If Juan were to be convicted, it would be by the testimony of the white inmates.

We were allowed to leave our rooms at mealtimes, but we were escorted by a dozen guards. They cleared the hallway of leprosy patients before they let us pass. Carville now posed a problem for the Bureau of Prisons. There was no way to secure the facility. Two doorways dividing the colony could be locked, but it was impossible to seal the breezeway and the corridors. If an inmate wanted access to the leprosy patients, it would be no problem. In the cafeteria, I nodded to Ricky and Chatto, my Mexican handball buddies, to let them know I was still their friend, but they just looked away. A few Mexican inmates stared into the eyes of those who had "ratted" and made stabbing gestures at their chests.

On the third night of lockdown, Link slipped out of his room to check on me and offer advice.

"You shoulda kept your job in the kitchen," Link said, "'cause you can fuck somebody up with boiling water." Link described how he once saw flesh melt off the skin of an inmate at another jail when his enemy had crept up behind him with a pot of water just off the stove.

"Socks is another thing," he said. Link described in detail the damage that could be inflicted by placing three or four padlocks inside a tube sock. He demonstrated how an attacker could swing the sock to build up momentum, the heavy locks inside gaining velocity on the outer perimeter like David's sling against Goliath. The impact would break ribs or fracture a skull.

"Thanks, Link," I said. "I'll keep that in mind."

"Now, if you don't wanna kill the motherfucker," he said, "just put bars of soap in the sock. It'll just bruise him and shit."

Link's advice, coupled with the fact that our rooms had no doors, wasn't comforting. He wasn't the only inmate making plans. A devout Catholic from Texas—a man who told me I was going to hell for taking Communion at the Catholic church—stole rat poison from the kitchen to use against any inmate who threatened him. Another man stole a baseball bat from the recreation department. Still another was planning to use his roommate's prosthetic leg to fight off intruders.

On the fifth day of lockdown, we heard the news.

The *Baton Rouge Advocate* reported that John Gray, fifty-eight, had died of a massive heart attack nearly four days after being stabbed by another inmate. A prison spokeswoman was quoted: "Gray received a superficial abdominal wound. It had nothing to do with his heart attack." She added that no other information could be released until the FBI completed its probe.

The man we had known for nine months was gone. And every inmate in Carville knew exactly what caused his death.

Lockdown was frustrating. No walking. No library. The leprosy patients, understandably, were nervous, too. We were never to pass a patient unescorted again. Guards were put at every corner of the inmate courtyard, and they increased the frequency of shakedowns and strip searches. One guard kept watch at the breezeway to make sure no inmates had any contact with the leprosy patients.

"I seen you talkin' to the old lady," he said to me. "Them days is over."

The guards had a light in their eyes I had not seen before, as if this were their first opportunity to act like real law enforcement officers. Running around in their SWAT shirts, they enjoyed the drama.

A week after the stabbing, as we were escorted to the cafeteria for breakfast, I heard a guard yell, "Against the wall!" Sixty of us parted like the waters of the Red Sea and stood with our backs to the walls. It was Ella. One guard walked in front of her, another behind. As she rolled slowly through us, Ella made eye contact with each inmate and smiled. When she reached the end of the line, where I was standing, I expected her to nod or wink or give me some kind of special sign, but she didn't.

I got the same as everyone else.

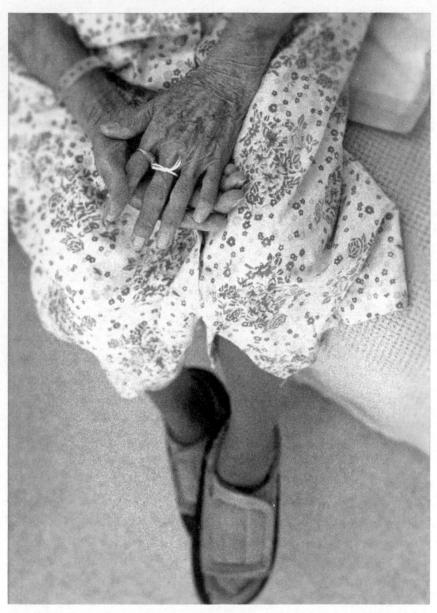

Betty Martin, who took furloughs from Carville to her home in New Orleans.

CHAPTER 66

In the midst of lockdown, I learned that my furlough had been approved. I would be released for five days. It could not have come at a better time.

In the administration building, behind the door marked R & D, I was frisked by a guard.

"You understand the rules?" he asked. I told him I had read the papers, but he reviewed them anyway. While on furlough, I was not to break any laws, leave New Orleans, use drugs or alcohol, go inside a bar, take prescription medication, go see a doctor, or eat food containing poppy seeds (apparently, it could cause a positive result on drug tests).

I signed my forms and the guard escorted me to the end of the hallway. He opened the door and said, "Be back by 8:00 P.M. Friday." And I walked out.

I still didn't understand the logic behind a furlough. Why would my captors, in the middle of my prison sentence, simply say, "Go home for a while"? But I was sure happy about it.

Mom waited in her small, maroon Isuzu with Neil and Maggie. On the drive to New Orleans, Neil and Maggie took turns sitting on my lap. We laughed and played games, and I pulled out a list I'd made of fun things to do during our time in New Orleans.

"Anything the two of you want to do," I said, "is great with me."

Mom's place in the French Quarter was built in the 1800s. The second-story apartment was long and narrow with fifteen-foot ceilings. A small balcony with wrought-iron railings overlooked Toulouse

Street. The windows ran from floor to ceiling. And when the bottom window was pushed up, I could walk under it without ducking. The back of the apartment connected to a huge wooden spiral staircase that led to the ground floor, where there was a slate courtyard with a small fountain and garden.

During my last few months of freedom, I had lived in the apartment above Mom's. Linda and the kids had lived there after I reported to prison, so Neil and Maggie felt right at home. The neighborhood was familiar territory for them. We walked to Le Marquis for fruit pastries. The manager of Le Marquis, a woman who grew up in New Jersey but donned a French accent for the tourists' sake, asked where I had been. I told her I had moved. With Neil and Maggie at my side, I didn't want to try to explain my absence. Maggie ordered a fruit tart, Neil got three donuts, and I ordered a croissant, although the poppy seed muffin looked enticing.

At Jackson Square, we tossed pennies into the fountain. Neil and I threw a Nerf football in a small patch of grass while Maggie ran in circles inside the gated park. We met my father there, who had come to visit. And my sister, Liz, who was living in New Orleans, introduced me to her fiancé, Sal. I didn't even ask her what excuse she'd given Sal about where I'd been.

Once everyone arrived, we toured the massive Catholic cathedral. We watched street performers walk on stilts and ride tall unicycles and perform acrobatic dances. One of the local magicians asked Maggie to assist him with a magic trick. A tall, thin contortionist had Little Neil squeeze into a tiny, transparent box. Then the man placed his six-foot frame inside. It was impressive.

Artists' booths lined the outer edge of the park's iron fence, where they sell paintings and charge $40 for a quick portrait. There were palm readers and fortune-tellers and voodoo ladies. A tarot card reader tried to solicit business from a group of young men walking by. "You can't change my future, old man," one yelled out. The tarot man yelled back, "No, but I can help you prepare for it."

At the end of the day, we all gathered around the granite base of the statue of Andrew Jackson on horseback and asked a passerby to take our picture. My first day of freedom, even though temporary, had been full.

Late that night, after Maggie and Neil fell asleep, I told Mom I was going for a walk. I roamed the French Quarter. I passed discos and zydeco bands. I looked through the window of M. S. Rau Antiques, a store so exclusive and expensive it could pass for a museum. Rau stocked rare commodes used by kings in eighteenth-century France, fancy diamonds that cost half a million dollars, two-hundred-year-old chairs designed by European royalty for easy access to multiple sex partners, globes that cost more than a year's salary, and contraptions designed by insane nineteenth-century scientists.

I continued down Royal Street and stopped at Hurwitz Mintz, a furniture store where Linda and I had spent $7,000 on two leather chairs and a leather couch. Before Carville, I'd had my eye on an expensive round conference table.

At the far end of Royal, I passed Mr. B's Bistro. Mr. B's had been one of my favorite restaurants. The staff had treated me like I was special because I published *New Orleans* magazine. On some evenings, I would spend more than $500 treating clients or friends to dinner. Now, I had less than $100 to my name. The warm yellow lights made the rich wood walls glow. I stood on the sidewalk and looked inside. I didn't worry about being noticed by anyone I knew. The people inside Mr. B's didn't bother watching the people outside.

I rounded the block and passed a couple of seedy strip joints. I wanted to step inside. Just to take a look. I felt like I might better relate to the patrons of the Artist Cafe strip club than any of my old hangouts, but I didn't want to violate the furlough rules. And I figured the bouncers wouldn't let me stay without buying an overpriced drink.

I walked back toward the heart of the French Quarter and thought about the patients at Carville. Some of the patients took furloughs, too, but only if a family member would accept responsibility. Before the 1950s, no more than ten patients could be away from the colony at a time. And many patients had been forsaken by their families.

I tried to imagine what it must have been like for Ella or Harry to go home after being locked away for so long. Betty Martin, the New Orleans socialite who contracted leprosy at age nineteen, took a furlough. She came back to New Orleans to visit her mother and father in their Uptown home, located just a few blocks from where Linda and I had lived. When Betty came back to New Orleans, she never left her parents' home. Friends and family, the few who knew the truth, came to visit her. She stayed inside; she didn't want anyone to see the woman who was now known by the alias Betty Martin.

I understood her fear, why she shrunk away from old contacts. I didn't want to mingle with the same people I had sought out before Carville.

As I meandered the French Quarter, I made a mental note to remember to climb the levee at Carville before I reported back from the furlough. I'd been so excited to see Neil and Maggie waiting in Mom's car, I'd completely forgotten to stop and see what direction the river flowed.

Before I went home for the night, I stopped at the A & P grocery at the corner of Royal and St. Peter. The store was tiny, aisles so narrow the store provided specially designed, miniature shopping carts. I bought Hot Pockets and applesauce and SpaghettiOs and macaroni and cheese. I stood in line behind a transvestite with a neck brace. In her heels, she was at least three inches taller than I, and she had a five o'clock shadow. She had two items in her basket: aspirin and cat food.

"Headache and the cat's hungry, huh?" I said. The man/woman looked at me, surprised, I think, that I spoke.

"Yeah," she said in a deep voice, "life's grand."

I turned and noticed the man behind me. He was morbidly obese; he had long, blond curly hair; and he carried a large, matted teddy bear. I recognized him. Before Carville, I had seen him walking Jackson Square. Sometimes he begged on street corners. Other times, I would pass him without making eye contact. I would make room on the sidewalk when he passed me so I wouldn't brush up against him or, in summer, get close enough to smell his sweat-soaked teddy.

I turned toward the man and smiled. He squinted his eyes, and

I nodded, the kind of nod you give someone you know, but not really. The man nodded back. He had a box of Cap'n Crunch under one arm.

"I love that cereal," I said.

The man glared at me for a moment and held tight to his huge teddy bear. "Me, too," he said.

As I waited in the checkout line, it registered that my place in the world had changed. I was perfectly happy to be right here—at the French Quarter A & P standing between the transvestite and the man with the stuffed bear.

CHAPTER 67

On the first two days of furlough, the kids and I rode go-carts, watched movies, jumped on trampolines, and went to the aquarium. I was determined to cram as much fun as possible into these five days, but there was one thing I needed to do on my own. Without Neil or Maggie or Mom or Dad. I wanted to travel to a remote spot in the city. A place I had never been.

As we walked back to our apartment, I asked a cabbie how much it would cost to get to the corner of Hagan and Perdido. He thought for a moment. "Won't cost you nothin'," he said, "'cause that intersection don't exist."

He had to be wrong, I thought.

"You got Hagan. And you got Perdido," he said, "but they don't never cross."

Late in the afternoon, I took the kids to the New Orleans public library, just down the street from the Loyola Street Post Office, where Jefferson had operated the X-ray machine. Neil and Maggie read children's books while I found a detailed city map. The cabbie was right. Hagan and Perdido did not cross. But once they had. Hagan, I discovered, had been renamed. The street formerly known as Hagan was now named Jefferson Davis Parkway, after the president of the Confederacy.

On the morning before I was to return to prison, I awoke just after dawn. I had arranged for Mom to watch the kids. The morning air in the French Quarter smelled of stale beer and horse manure. A night manager was hosing down the sidewalk in front of his club on Chartres Street. The Stage Door bar was still open. Five or six men sipped one last drink before facing the day. I walked to the edge of the Quarter, where I found a taxi.

"I have a strange request," I said to the driver.

He shrugged as if there were nothing I could possibly request that he hadn't seen before.

"I'd like to go to the intersection of Perdido and Jefferson Davis."

The cabbie told me to get in, and we drove down Canal Street. A few homeless men sought shelter inside abandoned retail entrances. Otherwise, the streets were empty.

The driver's hair was oily. He had the heat blaring, and he smelled like he had been working all night. I could only imagine the fares he might have picked up on any given evening in New Orleans. Businessmen returning from Bourbon Street, prostitutes and strippers, jazz musicians who played the Quarter but couldn't afford to live there.

We drove under the interstate to Broad Street and eventually made it to Perdido. The street ran under a raised highway. I had no idea what kind of neighborhood we might be entering. Perdido was filled with potholes and crumbling asphalt. We passed old warehouses and a few abandoned vehicles. One car was left in the middle of the street, as if the owner simply got out and walked away when the car stopped running.

Daylight had arrived by the time the driver pulled to the side of the road. We had traveled more than two miles on Perdido. He checked the meter. "Eight dollars," he said.

I asked if he could wait. I needed only five minutes.

I stood in the middle of the intersection. On one corner, a dozen telephone repair trucks were parked in a lot protected by a chain-link fence with barbed wire on top. Across the street, metal pipes, probably left over from a public works project, were stacked under an overpass. A sign read NO DUMPING, $100 FINE. It could have been any corner of any city, but one hundred years earlier a two-story wooden cottage sat at this intersection. The building had been rented by the city of New Orleans. Local papers called it a pesthouse. In 1894, about a dozen men and women with leprosy lived here.

During that year, as the result of coverage from the *Times-Picayune*, citizens demanded removal of the occupants to prevent an outbreak of the dreaded disease. A group of men had anonymously threatened to burn the house and all who dwelled within.

A physician at Tulane Medical School had wanted to establish a leprosy hospital in the city, but fate intervened. The public's fear of exposure made New Orleans less than suitable for a leprosarium. The physician persuaded a friend, who was also a member of the Louisiana legislature, to lease thirty acres of land a little over an hour north of the city to establish a colony. But in order to get the land, the legislator had to lie. He acquired the lease under the auspices of establishing an ostrich farm. On the night of November 30, 1894, five men and two women—who were never told of their destination—were transported eighty miles up the Mississippi River. They made passage on a coal barge. (It was against the law for lepers to use public transportation.) At dawn, the seven were unloaded at the plantation where they would live out the rest of their lives.

Nothing of the pesthouse remained. I stared at the parking lot and tried to imagine the home. I tried to picture the victims of the disease, as well as the physicians who hurried them out of the city. A century separated us, but I felt a deep connection to the men and women who had lived in the pesthouse. At Carville, we had found a place of refuge.

On the ride back to my mother's apartment, I told the driver about the story of the nondescript intersection and its historic significance in the establishment of the only leprosarium in the continental United States.

"How you know so much about it?" he asked.

"I live there," I said. "Upriver."

For the first time he looked attentive. He stared at my reflection in the rearview mirror like he might be worried.

"Five-day pass," I added, happy to not be just another boring fare. And proud to be a part of Carville.

At the end of our short ride together, I would tell him that leprosy wasn't much of a threat, and that I was just an inmate on furlough. And I would probably give him a bigger tip than he deserved because deep down, I suppose, I still cared too much what people thought about me.

CHAPTER 68

After a dinner of gumbo and cornbread, Mom drove me back to Carville. We hit the levee and turned right on River Road. The last time I had been on this road, on my way to prison, I had no idea what lay ahead. I thought I was headed to an ordinary federal prison. I had no idea of the absurdity, complexity, tragedy, and magic that was Carville.

Just before dark, we passed the lights of the chemical plants, the Carville family country store, and the road sign that read PAVEMENT ENDS TWO MILES. Then we passed a lone, ancient oak tree. The last one before the colony walls.

I couldn't help but think about my friends who had taken this same path. Ella on wagon. Jimmy Harris in his father's Buick. As we approached the colony, I noticed a landing for ferries on the Mississippi River. In 1894, as reported in *The Star*, Fritz Carville (grandfather of the Clinton aide James Carville) rode his pony to this spot. The ten-year-old boy, accompanied by a farmhand, had heard about the new ostrich farm. Young Carville watched the barge dock on the riverbank. But instead of unloading ostriches, seven leprosy patients were left at the abandoned plantation. The farmhand looked at the boy and said, "Little Boss, them ain't ostriches—thems sick folks!"

As Mom drove toward the prison gate and the sun disappeared behind the levee, I sat between Neil and Maggie in the backseat. And we held hands.

PART V

Spring

CHAPTER 69

Back inside the colony, a guard gave me a urinalysis test. Another one performed a strip search. As I bent over, I realized I had, again, forgotten to climb the levee to see if the river actually flowed north.

Contraband- and drug-free, I was cleared to enter the prison. The guards did not escort me, which meant lockdown was over. I was still anxious, concerned about tension between the white and Hispanic inmates over the stabbing. In the hallway, I saw prisoners waiting in line for the telephones and inmates darting into the TV rooms.

Smolkie, an inmate from New Orleans who loved to gossip, yelled, "Hey, Clark. Have you heard? They sendin' everybody home!"

"What?"

"They closin' da place!"

"Why?" I asked.

Smolkie shrugged. "The stabbin', the smugglin', who knows? But ain't nobody gonna be here much longer."

None of the inmates knew what had happened, but Smolkie was right—the prison side of the colony was closing down. Any inmate who had more than six months remaining on his sentence would be transferred to another prison. Those with fewer than six months, including me, would be considered for halfway house, or early release. Whatever tension lingered from the stabbing had been wiped away by prospects of early release. The Mexicans played handball with the white inmates. Ricky and my other friends waved and smiled at me again. The inmates were downright overjoyed, including Link.

"What you did out there, Clark Kent?" Link asked.

"Spent time with my kids," I said.

"No damn parties?"

I shook my head.

"Goddamn," Link said, "they let me out for a week, I'm gonna get me a limousine, some crack, and some women."

With the prison closing, the leprosy patients were free to stay. Ella and Harry were happy again. The patients had regained control of their home. The Bureau of Prisons had completely abandoned its plans. Rumor was the Bureau administration wanted to avoid the risk of another stabbing. Or perhaps they realized that if Smeltzer could help smuggle muffulettas inside for forty prisoners, the prospects of keeping out guns and drugs seemed virtually impossible. But no one knew for sure.

The week before Mardi Gras, our questions were answered. An official statement released by the Bureau of Prisons appeared in the *Baton Rouge Advocate*. The Bureau claimed it couldn't make the necessary physical changes deemed vital to secure a prison. The reason: Carville had recently been listed on the National Register of Historic Places. With the historic designation, the renovations the warden planned would be impossible.

The closing was a victory for the patients over the Bureau of Prisons. The warden loved Carville and authorized millions of dollars in improvements. In his mind, the colony was perfect for a huge prison, isolated, a white elephant for anyone else. The Bureau had the funds to convert the facility into something spectacular.

The warden had grand plans, but he had not anticipated the response of the residents or the persistence of the ladies of the Carville Historical Society, who ultimately prevented his dreams.

Smeltzer gloated in the hallway. "I knew we'd beat you!"

"We hate the Bureau, too," I said.

Smeltzer looked confused.

"Well," he said, "you've been nothing but trouble since you got here."

Smeltzer was about the only patient I didn't like. "How are you going to make money with us gone?" I jabbed back.

He stared at me for a moment. "Just wait for the Mardi Gras parade!" he said.

"What's going to happen?" I asked.

"We had to change the route because of you people," he said. "But we got something for you."

The Carville Mardi Gras parade had always been one of the year's biggest events for the residents. Since the 1920s the patients built floats, decorated their wheelchairs, and donned elaborate costumes for the festival. They paraded through the entire colony tossing doubloons and beads. But this year would be different. Because of the stabbing, we were told the patient parade route would be limited to a small hallway and the recreation room. In addition to redirecting the route, we heard that this year's visiting list would be restricted.

As part of the closure, the warden told the guards they would need to find new jobs. Consumed with their own futures, they lost interest in us. The Bureau promised to help them find jobs at other prisons, but the river region was dotted with petrochemical plants with higher-paying jobs. So the guards worked on their résumés, cover letters, and job applications.

One of the guards remembered the flyer I had posted for the résumé-writing seminar and, during study hours at the education department, asked if I would review his. I edited his résumé for continuity, verb-subject agreement, passive voice, and basic sentence structure. I marked suggestions for spacing and organization.

The next day, three more guards brought in résumés. I suggested a complete rewrite on one. With another, I recommended a format change from narrative to short bullet points. Soon I was proofing for a dozen guards. One asked if I would review his automobile lease.

Jefferson suggested I intentionally insert errors, but I felt pretty good about lending a hand, even to guards.

I reviewed a cover letter addressed to Harvey Press in New Orleans. Ms. Carter, a secretary in the education department, had applied for an administrative assistant position. I told her I had done a good bit of business with Harvey Press, I knew the owners, and I would be happy to write a letter on her behalf, if she'd like.

I'd never seen anyone so happy to have a federal convict's recommendation. I borrowed her typewriter and hammered out a fairly glowing endorsement. Ms. Carter reviewed the letter and gave me a big hug. Her eyes went wet. "You good people, Mr. White," she said.

The white-collar inmates were scrambling to find work, too. They pored over the *Wall Street Journal*, *BusinessWeek*, and *Forbes*, scouring the pages for opportunities.

Doc met with a wealthy inmate about investing in his impotence cure. Dan Duchaine was finishing his book. Frank Ragano was planning his book-signing tour. My friend Danny Coates had developed plans for offshore gaming.

An inmate called Super Dave wrote a business plan for a telephone scam. The plan outlined details like hiring unwitting young women to get investors to pay $99 for the opportunity to make tenfold returns. The plan *actually* described the characteristics to look for in hires so you stay in town for two weeks, pack up, and never show your face again.

"Clark Kent!" Jefferson said as he danced into my room. "I'm getting out today! And I'm going legit!" Jefferson had saved over $40,000 from the cash he'd taken from birthday and holiday and graduation cards. Though he'd never been caught for his X-ray machine work, he couldn't be rehired by the post office.

Jefferson said he was opening a business with his sister. They had put in an application at a bona fide franchise company, using the seed money Jefferson had taken from the X-ray machine. I stood and shook his hand. I told him I was proud of his new, legal approach to business. He smiled like he always did and started dancing down the hall.

"Hey, Jefferson," I called out, "what franchise are you going with?"

Jefferson looked over his shoulder and grinned, his gold teeth sparkling. "Mail Boxes, Etc.," he said.

. . .

In the days when leprosy patients were quarantined in the United States, the rationale for confinement was public welfare. It was widely believed leprosy patients would spread a scourge on society. For decades, men and women who had done no wrong were imprisoned for the public good.

As I listened to the inmates' schemes to reenter the world, I did not miss the irony that we were being released while the innocent remained behind. We were the scourge on society. We were the "lepers."

And we were about to be set free.

CHAPTER 70

Not to be left out, I drafted my own short business plans—five streamlined plans, including marketing strategies, product plans, and financial projections. Three plans called for magazine launches: *Southern CEO*, for the business decision makers in the South; *Slammer*, a magazine for prisoners; and a yet-to-be-titled magazine for college-bound students. I also wrote a plan for a boat rental business at Walloon Lake, as well as a prospectus for leading seminars.

The prospect of operating a new business was exciting, but I was also apprehensive. I had no seed money; borrowing was out of the question. Taking on investors opened the door to financial loss for others—a risk I would never take again.

As I reviewed the plans, I was interrupted by a guard. My father had come to visit.

Dad had recently moved to Alexandria, Louisiana, to take a position as a federal administrative judge, so a trip to Carville was just over two hours. He greeted me with a hug, and we sat at a table.

"Drove through Gross Tete, today," he said with a grin. Dad loved the names of Louisiana towns. He found joy in little things.

Dad had visited whenever he could, and I felt lucky that he came so often. He always introduced himself to my inmate friends. He was friendly to the other families. He didn't lecture me; he didn't preach to me or scold me or warn me about future actions; he just wanted to be with his son. Even in a prison visiting room.

His relationship with his father, our namesake, Neil White Sr., was different. His dad was a drinker, too. His nickname was the Old

Dog. As anyone who grew up on the Mississippi coast could tell you, the Old Dog was the life of the party. And he didn't miss many. After he eloped with my grandmother Martha, he never needed to work. He had a sharp mind and was offered a job in the White & White law firm making $6,000 per year, but my grandmother's interest income was more than $50,000 per year. In lieu of work, he drank. He turned his wit into a wicked tool to mock those who were too earnest, or not part of his entourage. The Old Dog didn't intend to hurt anyone. He would probably say he simply didn't want to be inconvenienced. But when you marry a woman at age eighteen and produce seven offspring, convenience and solitude are hard to come by. His frustration, coupled with too much drink, sometimes came across as cruel. His edge was shrouded in humor, but it was most costly to the people who loved him—his wife and his children.

But my father didn't deal with his father's cruelty by passing it on to the next generation. He bore the full weight of it himself. He didn't lash out. He never made fun of us. To his children he was loving and loyal and sweet. There was not a mean bone in his being.

When I was young, I considered him weak. I thought he stayed in bed too much. I chalked it up to someone who just couldn't cut it. I wanted to do the opposite. I wanted to make a difference, leave a legacy, be the kind of man who was admired, and return glory to the family name.

Back then, I didn't understand. And until recently, I had been too absorbed in my own ambition to fully appreciate how strong he was.

As a sober man for the past fifteen years, my father had spent his days helping others get sober. One person at a time. He did it quietly. Anonymously. And for the last year he had spent many a weekend driving to a remote colony in South Louisiana to sit by his son in a prison visiting room.

The men I had admired so much, the men who were so different from my father, would never consider sitting with me now.

My father put his arm around me. He wanted me to know I was loved. And for the first time, I realized I wanted to be more like him.

CHAPTER 71

The prison population dwindled. U.S. marshal buses and vans arrived daily to transfer inmates.

On a sunny spring afternoon, as I returned from the education department, I saw Link. He was escorted by two U.S. marshals. They had restrained him in handcuffs and leg shackles.

When I reached Link, I asked the marshals if I could say good-bye. They told me to make it quick. Link wore an orange jumpsuit and his signature smile.

I held out my hand, but his hands were bound by the chain around his waist.

"I never got your address," I said.

He looked at me and laughed. "You think we gonna be friends on the outside?"

"Sure," I said. "We're friends now."

"You don't know shit, do you?" One of the marshals snorted, but I didn't care. Link had helped make prison tolerable for me, and I did consider him a friend.

"So we're not going to be friends anymore?"

Link shook his head like he knew something he couldn't explain to me.

The marshals grabbed his arms and walked him down the hallway toward Receiving and Discharge. I stood in the middle of the hall and watched. They turned to go inside, and Link leaned his head back as far as he could and let out a big laugh. The marshals jerked him up and dragged him through the door. I expected to hear him yell out one more time about how boring or white or stupid I was, but he didn't. I heard the door lock behind them. And the prison was quiet.

A Carville patient Mardi Gras parade, ca. 1950s.

CHAPTER 72

In preparation for the annual patient Mardi Gras parade, the leprosy patients and staff built floats out of wheelchairs and bicycles and carts. The parade route was short, but the patients took it seriously. They even minted their own doubloons. The gold, purple, and green coins featured the patient recreation hall on one side. An armadillo graced the back side of the doubloon.

Armadillos, it was discovered in 1971, were the only other creature to naturally contract leprosy. In some parts of Texas and Louisiana up to 20 percent of wild armadillos are infected with the disease. Some experts believe that they comprise the greatest reservoir of *M. leprae*, and many suspect leprosy can be attributed to handling the animals.

Since the bacteria had never been replicated in a test tube, the armadillo accelerated the discovery of new treatments. After 1971, leprosy patients were no longer the guinea pigs; that role was passed along to the nine-banded armadillo. The patients adopted the animal as their informal mascot.

For this year's parade, the patients had built a special float—the "thing" Smeltzer had mentioned earlier—specifically for the Bureau of Prisons. Chase and Lonnie, the inmate trusties who could go anywhere on the patient side, anytime, helped the patients build it. The float depicted a huge tombstone. Engraved in the faux marble were the letters "RIP, BOP."

The warden had heard about the Mardi Gras float and he was livid. After losing the battle for Carville, the float was like rubbing salt in a wound.

The warden wrote a letter to Dr. Jacobson, the Public Health Service officer in charge of the leprosy patients, demanding that the float be removed from the parade. But what the warden didn't understand, Dr. Jacobson did. Mardi Gras was a playful, festive time. And he refused to interfere with the patients' plans.

On the day of the celebration, hundreds of inmates lined up at the corridor to watch the patients parade, but we couldn't see very well since the parade wouldn't proceed down the breezeway. The patients couldn't throw us any beads or doubloons because the windows were covered with screens, but we could hear the band leading the procession. And the patients could hear us cheering for them.

I caught a glimpse of the tops of wheelchair floats. As they rolled by, I tried to identify my friends. But most of them wore masks.

As the floats moved toward the recreation hall, the patients waved to the inmates. We were their only audience. The warden watched from the large glass window on the second floor of the prison side. He stood with his arms crossed, his two lieutenants at his side.

When the float with the tombstone turned the corner, Chase gave us a signal, and more than a hundred inmates cheered and jumped and danced like we were standing on Bourbon Street. A roar that sounded like a riot echoed through the inmate courtyard.

The warden had been beaten by an unlikely convergence of leprosy patients, nuns, convicts, and the men and women who worked to get Carville on the historic registry. It was time to celebrate.

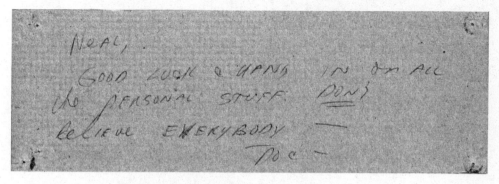

The note that Doc left me when he was transferred.

CHAPTER 73

Five days after the Mardi Gras parade, on Ash Wednesday, I knelt at the altar of the Catholic church. I hadn't decided what to give up for Lent. And I wondered about the six or seven leprosy patients to my left. What would a man or woman who had lost so much forfeit for the Lenten season?

After the service, I caught Ella in the hallway.

"What are you giving up for Lent?" I asked.

"Hopscotch," she said. Ella gave me a smile. "You?" she asked.

"Freedom," I told her.

But I didn't really know what to sacrifice. Most of my temptations were already beyond my reach: wine, money, cars, clothes, houses, boats, vacations, and fine cuisine.

When I returned to my room, Doc's bed was stripped. His locker was empty. I opened the closet. His journals were gone. The marshals had transferred Doc, and I hadn't had a chance to say good-bye.

The room felt empty. I would miss him—our conversations, his theories and musings, our time together.

One by one my inmate friends were leaving. I felt alone, and oddly sentimental about the friends I might never see again.

That night, just before the ten o'clock count, I found a note Doc had left under my pillow. The short message was scribbled on a brown paper towel. It said:

Neil

Good luck and hang in on all the personal stuff. Don't believe every-body.

Doc

CHAPTER 74

"If you're not careful," Jimmy Harris said while riding his three-wheeled bike down the hallway, "we're gonna outnumber you." Jimmy was right. The inmate population was down to about 150. During the last few weeks, over 300 inmates had been transferred or released. In my unit, Dutchtown, empty beds outnumbered inmates.

On a Sunday afternoon, the guards announced that three of the seven inmate dorms would close. The guards were so disinterested they didn't even bother to make room assignments. They told the inmates in the three mothballed units to find an empty bed, anywhere.

In prison, a roommate can make your daily life tolerable or miserable. Within minutes, more than forty inmates, holding as many clothes and personal effects as possible, dashed to find a room. I imagined it being not much different from the 1800s when men raced to stake claim to land in the territories of the West. Dan Duchaine, out of breath, ran into my room and tossed an armload of clothes on the bed formerly occupied by Doc.

"They're closing St. Amant," he said. "I claim this spot." I nodded, and Dan left to gather what was left in his old room. I was flattered that he wanted to be my roommate. Dan's inmate guinea pigs would have given just about anything to room with the Guru. Dan seemed a good replacement for Doc. Same temperament. Same smarts. And the conversation would certainly be rich.

$\bullet \quad \bullet \quad \bullet$

During Lent, I established a routine. I spent my mornings in the education department. Six or seven of my students had already passed the GED test, and I was making progress with the other inmates. I felt good. This was just the sort of thing I wanted to do when I was released—help people, put my time to good use.

I spent my afternoons and evenings walking the track. Asking myself how would I act toward those who shunned me. Asking if I could remember to live simply. Asking if I could repay my victims without risking another financial failure. And asking how I would provide for Neil and Maggie while meeting my other obligations.

I thought of all I had taken for granted. I'd had the support of parents and family and employees and bankers and investors and friends. So many people wanted me to succeed. So many people would have helped me had I asked. So many people supported me financially.

But just like my grandparents, Neil and Martha, who were young and talented and could have done so much good, I had taken my blessings for granted. I had thrown them all away.

And when I wasn't thinking about how I could change, I thought about the friends I'd made at Carville. I was grateful to know them all. I was already starting to miss them. I was in serious danger of becoming overly sentimental. Especially when it came to Ella. I was acutely aware that we would soon be separated. And I was afraid I'd never find another friend like her.

On a Saturday morning, I stood in the breezeway entrance and watched Ella crank her wheelchair toward the patient side. Even after fifty years she had not perfected a synchronized crank. Standing behind her, I saw her waver. She would veer off a bit to the right and then adjust with a longer crank to the left. She constantly adjusted her course. And as long as she paid attention, she would never hit the corridor wall. Ella veered and corrected, veered and corrected, a thousand times a day.

While I was trying to live a quiet Lenten season, the leprosy patients and the PHS staff were planning a fall celebration—Carville's one hundredth anniversary—commemorating the arrival of the first seven patients in 1894.

The plans had almost been finalized. For the patients, the event served a dual purpose—a centennial observance, as well as a celebration of the exodus of the inmates. James Carville, Bill Clinton's adviser, was scheduled to make a speech. Carville's family had made a fortune selling goods to the colony. Other events included an open house, special exhibits, a golf tournament, the premiere of *Exiles in Our Own Country* (a movie about the patients at Carville), the unveiling of a mural wall, special tours, speeches by dignitaries, the publication of a centennial history book, and speeches by patients and activists from around the world. Jimmy Harris had been commissioned to create an oil painting depicting the landing in 1894. And Louisiana governor Edwin Edwards had agreed to endorse an official State of Louisiana resolution recognizing the one hundredth anniversary event.

I wanted to attend the centennial celebration. My growing devotion to Carville—its history and the leprosy patients—made me not want to miss anything. But I knew I wouldn't be invited. The Public Health Services, after its skirmish with the Bureau of Prisons, considered the prison and the inmates a blemish on its history. But I felt proud to live in a room that offered a century of safety for leprosy patients. I was honored to take Communion in the same sanctuary where society's outcasts asked God to console their suffering. I felt privileged to live and work and play in a place that few had ever seen. And I was grateful I had been imprisoned here, in a leprosarium, where I could begin to rebuild my life in a different way.

CHAPTER 75

On a bright day in April, Dan Duchaine yelled out, "Smeltzer's got a prostitute!" Dan didn't get excited often, but this set him off. With the proceeds from the pork chops and newspapers and muffulettas and pedicures, Smeltzer had bought a hooker for the patients' spring dance.

"Do you think he paid her in quarters?" I asked.

"Just imagine the conversation at the brothel," Duchaine said, mocking the lady of the house: "C'mon, girls, it's leper day."

Smeltzer wasn't the only one with a date. Spring, and the prospect of freedom, brought romance to the colony. An investment banker from Texas alternated weekend visits between his wife and his girl-friend. Another inmate in his seventies found a lady friend through personal ads. Father Reynolds performed a wedding in the Catholic church between an inmate named Wes and his fiancée, a free woman. CeeCee upped her attempts to woo a new lover by offering unlim-ited maid service. And the patients were preparing for the dance. The women patients were getting their hair styled and ironing their dresses. The men pulled out their best clothes and filled their flasks with whiskey.

In the early days of the colony, male and female patients had been segregated. Living quarters were separate. Men and women ate at separate times. Mingling was an offense punishable by time in the colony jail.

But everything changed when Dr. Denny arrived in 1921. He came from Culion, a leper colony in the Philippines, and assumed the role of director of the national leprosarium. He had seen patients suffer. Segregation from society was enough, he believed. Segregation of the sexes was doubly cruel. Patients still weren't allowed to marry, but Denny dismantled the fence that kept the sexes apart. The year he arrived, the patients held their first dance.

"Be my date?" Ella asked the afternoon before the dance. Ella smiled. She knew inmates weren't allowed at social events.

"Wish I could," I said.

Ella said she couldn't talk long. She had much to do to get ready, and she didn't want to miss the beginning of *Matlock*.

I told her I couldn't wait to hear about the dance.

After dinner, Dan and I returned to our room. A guard stuck his head in the doorway.

"White! Duchaine!" he yelled. "Report to the patient ballroom." Dan and I were instructed to help set up tables and chairs and unload equipment for the dance. I understood why they selected me; I was work cadre, but Dan, who had suffered a stroke, was in khaki. A medical inmate. But he didn't seem to care.

Around 7:00 P.M., the patients rolled into the decorated ballroom. Smeltzer arrived with his date. A woman in her late forties with en-hanced breasts that caused a slight curve in her back, like they were too heavy for her frame, she did look like a prostitute. She wore a low-cut red dress with tiny straps that disappeared into her shoulders. Smeltzer escorted her to a table without the benefit of his walker.

Smeltzer gave the woman a drink and held his own with both hands. His fingers were tight nubs that looked ready to burst through the skin. He put the drink to his lips and turned it up.

Ella cranked across the scuffed ballroom floor. Her hair was down and neatly combed. Her brass hoop earrings swayed with each turn of her hand. She stopped and smiled at me. Her skirt was empty and flat and draped over the edge of her wheelchair.

"Gonna put your legs on?" I asked.

She shook her head. "You gonna stay for the first dance?"

An old hippie tapped the microphone and said something about how much he loved the Carville gig. A loud, off-key rock song pumped out of the speakers. The patients limped and wheeled and slid their walkers out onto the dance floor. Stan, the blind jazz musician, scatted in the corner with his white tapping stick—great rhythm, head and mouth cool like Ray Charles. Harry shuffled to and fro. Smeltzer held his hands over his head and shook them toward heaven like this would be the last song he'd ever hear.

Arms flailed. Bandaged hands flew into the air. Whiskey spilled on the floor.

Ella swiveled her wheelchair to the beat in tiny movements. She motioned for me and offered her hand. Her long, elegant fingers, well cared for despite dead nerves, were soft and smooth. She swayed inside her chair. I took her hand and matched her rhythm. Harry, from under his hat, gave me a nod. I moved behind Ella and leaned her wheelchair back. I pushed her chair around the edge of the ballroom and turned her in circles. She stretched out her arms like she was flying. She looked over her shoulder at me and smiled. Her eyes were wide and alive like I imagined they had been when she ran and danced as a reckless young girl. We spun and twirled and slid until we were dizzy and the room disappeared.

Toward the end of the song that lasted too long, Chase and Lonnie, the trusty inmates who spent more time than any others helping patients, danced into the room from the leprosy side. Chase was tall and fit from Duchaine's diet and exercise program. He wore a cap to keep his long black hair out of his eyes. He danced into the middle of the patients. Chase tapped the shoulder of the prostitute and got a little too close for Smeltzer's liking. Chase was a good dancer, and the woman liked the attention. She laughed and danced lower and lower as Chase matched her moves.

The music stopped. The patients clapped as best they could.

Duchaine looked at me like I was demented, but I didn't care. I pushed Ella back to the center of the floor. I took her hand and bowed.

"What the hell are you doing!?" Smeltzer yelled.

I looked up. He was screaming at Chase, who stood perfectly still. The room was quiet, and I was glad not to be the focus of Smeltzer's ire.

"You're not invited!" Smeltzer said, pointing the remains of his index finger at Chase. Then he turned toward me. "You either! No inmates at our party!"

I looked for my friends. Ella's smile had disappeared. Harry looked at the floor.

"Go on," Smeltzer said to Chase. "You're not welcome here." Then he looked at me. "Go on. Get!"

Chase and Lonnie scurried away together. Dan and I walked behind them along the concrete corridor that led back to the inmate wing.

"Jesus," Dan said, "did we just get kicked out of a leper dance?"

UNITED STATES GOVERNMENT
MEMORANDUM
FEDERAL CORRECTIONAL INSTITUTION
CARVILLE, LA 70721

DATE: April 7, 1994

REPLY TO
ATTN OF: Alice Caronia, Unit Secretary

SUBJECT: WHITE, Neil W. III
Reg. No. 03290-043

TO: All Concerned

The above referenced inmate is scheduled to be released via CCC Furlough Transfer from FMC-Carville on Monday, April 25, 1994 at 8: 00 a.m. The travel itinerary is as follows:

DEPART: FMC, Carville via family member on 4-25-94, at 8:00 AM

ARRIVE: Volunteers of America in New Orleans **NO LATER THAN 9:30 AM.**

Distribution: R & D
File
Inmate
Lieutenant's Office - Erin
Halfway House
Business Office

My release notice.

CHAPTER 76

Late in the evening after the dance, the guards came for Chase and Lonnie. They were escorted to the hole. Rumor was they would lose their good time. If they came back for me, and I lost the fifty-four days of good time I had earned, I might be transferred to another prison, instead of released.

When Mr. Flowers arrived in my room, I assumed he had come to take me away. I was prepared to confess to dancing with Ella. I had already promised myself I would tell the truth, no matter the consequences.

"Congratulations," he said, as he handed me my release papers. I would leave Carville on April 25, 1994. I would report to a halfway house on Magazine Street in New Orleans in a matter of days.

I was relieved that the guards never came for me. Perhaps they didn't consider my dancing a violation since I was in the ballroom under orders. Or maybe because I didn't make a move on Smeltzer's prostitute.

I lay in bed after lights-out, the time of night when, as a child, I would dream about being on the pages of *The Guinness Book of World Records*. I should have been thinking about my release. About ways to remember the lessons I'd learned. But as much as I tried to push the thoughts of breaking records out of my head, it didn't work.

I had always wanted to achieve one feat never before attained by man, and I couldn't help but imagine that it had happened. Tonight. The first ever to be ejected from a leprosy dance.

A garden lined with Coke bottles, ca. 1950s. The local Coca-Cola distributor had refused to accept the returnable bottles from the leprosarium.

CHAPTER 77

I stood in the breezeway and waited for Ella. I had two things to tell her. First, I would be going home in a matter of days. The second was more difficult. With my release looming, I was acutely aware that I had not really changed during my year at Carville. I had *decided* I needed to change, but I was the still the same man who walked through the gates a year earlier. I awoke each morning wanting to do something great. I wanted to set records, whether it was the most magazines sold or a 100 percent success rate with my GED students. I relished accolades, even if prompted by a fruit and vegetable garnish. And, clearly, I had not abandoned my dream of being first, even if that first was being evicted from a dance.

"Hard on yourself," she said, after I told her my apprehension.

I shook my head. "Everybody says I need to become a new person before I get out."

"You is what you is." Ella took a deep breath and looked across the inmate courtyard. "You know 'bout them drink bottles?" she asked.

"No."

Ella intertwined her fingers like she always did when she told a story. In the early days of Carville, she explained, the Coca-Cola distributor from Baton Rouge sent chipped and cracked Coke bottles to the colony, so he could refuse to accept the return bottles. He feared a public boycott if customers discovered the glass containers had been touched by the lips of leprosy patients.

"More drink bottles than you ever seen," she said. The crates of bottles filled closets and storerooms. But the patients discovered new uses for the nonreturnable bottles. They used them as flower vases with beautiful arrangements. They became sugar dispensers in the cafeteria. For impromptu bowling games on the lawn, the bottles were used as pins. They were turned upside down and stuffed into the dirt to line flower beds and walks on the Carville grounds.

"CoCola bottle still a CoCola bottle," Ella said. "Just found 'em a new purpose."

CHAPTER 78

The day before I was released, I packed my belongings. Everything I owned fit into two boxes. I couldn't believe a year had nearly passed.

I thought about my conversations with Ella, conversations I would revisit for a lifetime. Of all the wonderful things she taught me—the importance of home, about what people think, about being with my children—the most important might be the story of the Coke bottles.

For five months, I had agonized over how I should change. I examined the details of my past, the character flaws that contributed to my personal failure, the allure that applause held for me, my discovery that a pristine image could cover dark secrets, my attempts to balance bad deeds with good, and my optimism unchecked by good financial sense.

But I knew my essence had not really changed. I would always be the same person. Same skills, same personality, same character traits.

The story of the Coke bottles was a wonderful parting gift. I didn't need to be a new person. I needed a new purpose. If I could follow Ella's lead—live simply, hide nothing, help others—maybe I would find a new purpose for my life. The challenge would be whether I could hold on to, and remember, the lessons when I lived on the outside.

Living simply might be the easiest. Many of my temptations would be out of reach. I'd never be asked to be on the board of directors of a bank. I'd never be asked to serve as treasurer of a club. I'd never be elected to the vestry of a church or be asked to head up the steward-ship committee.

Hiding nothing would be a struggle. It went against my nature. But Ella was a great model. If I could embrace my criminal convic-tion, if I could be transparent about my scars and indebtedness, not hide them, just like Ella did with her leprosy, it would be a step in the right direction. I even had a few things working in my favor. The move to Oxford, one I had dreaded, seemed to fit. In a large city, my felony conviction would be easy to hide. But in Oxford, I would have no choice. I couldn't hide my past even if I wanted to. And that was good.

I didn't know exactly how to go about helping others. But if I could remember that truly great acts are the small, quiet ones that no one hears about, that would be a start. I could look for ways to help people in need of a boost, to align myself with underdogs. I needed to remember Ella and Harry. Their intent. Perhaps it didn't matter what I did to earn a living, as long as the motive was to help others and not just gain attention.

I liked that the bottles were chipped and broken. They were dam-aged goods. Nonreturnable. I felt the same way. I could never go back to the place I'd been. I could never regain my reputation and cred-ibility. I would never have a flawless image.

As I started packing my boxes, I remembered the day I was sen-tenced. When Judge Walter Gex said, "Eighteen months in federal prison," I couldn't believe what I was hearing. After imposing my sentence, he warned me about living as a convicted felon and recited a long list of restrictions. I didn't remember much of it. But I remem-bered his parting words. As he left the courtroom, he looked at me as if he had been troubled by his decision.

"Neil," he said, "I hope you can make something good come out of this."

CHAPTER 79

My last night as a federal prisoner, a few inmates organized a party in an empty room in the Dutchtown unit. They all pitched in and made a vat of instant soup with slices of summer sausage from the commissary. Brownies from the vending machine were spread out on paper towels. Larry played his fiddle, the one tune he knew, and we recounted stories of the last year. We laughed about Smeltzer's muffuletta scheme and his prostitute. Slim told about looking for "cubicle hairs" in the women's restroom. I smiled whenever someone reminisced about Link, Doc, CeeCee, Frank Ragano, or Ms. Woodsen's butt.

We were the last of the inmates. We would all be leaving soon. We'd had some fun together, even in prison. The party didn't feel much different from the last night of summer camp sitting around a fire telling stories about the session. But we wouldn't see each other next summer.

At 9:00 P.M., a guard told us to shut the party down. Brady passed out small sheets of paper so we could exchange addresses and phone numbers. Gary reminded us that these would soon be obsolete since we'd be communicating over something called the World Wide Web, but I was skeptical.

On my way back to my room, I checked the Call-Out sheet again. April 25, 1994. Neil White. Receiving and Discharge. 8:00 A.M.

I emptied my locker and left my boots outside the door for anyone who might need them. I gave my *Playboy*s to my friend Danny and a long-sleeve T-shirt to Sergio. Then I packed the rest of my belongings: a few pairs of socks, six T-shirts, five books, photographs, and a few remaining letters and notebooks. Everything fit into two cardboard boxes.

I set my alarm for 7:00 A.M. and climbed into my prison bunk for the very last time. I put my arm over my eyes to block the light and waited for sleep to come.

PART VI

My Last Day
April 25, 1994

CHAPTER 80

I dropped my boxes at Receiving and Discharge, in the same room where I was first strip-searched nearly a year ago. Then I walked to the cafeteria to say good-bye to the kitchen staff. I wanted to go into the patient dining hall to say good-bye to my friends on the leprosy side, but I didn't want to risk breaking the rules on my last day. I did get a glimpse of a few patients through the lattice wall, but I couldn't get their attention.

After a breakfast of french toast and sausage, I made my way around the colony for a few farewells. I'd arranged to meet Ella and Harry in the breezeway a few minutes before eight to say our good-byes. On my way, I encountered Father Reynolds and reminded him it was my last day. He stepped off his bike, put his hands on my head, and said a short blessing. Then he said in his soft, stammering voice that I was welcome in the Catholic church anytime.

At the education department I said good-bye to the six or seven inmates who still attended class. I thanked Patty, the librarian, for her efforts, especially for forming the book club. Ms. Woodsen stuck her head through the library door and interrupted. "You leaving today, Mr. White?" she asked.

"Yes, ma'am," I said.

"You just might have something comin'." She smiled. I think Ms. Woodsen liked me. She seemed sincere in her good wishes, and I felt bad about laughing at the jokes about her rear.

Ms. Carter, the education secretary, started to cry when I entered her office. "I'm sorry," she said, sniffling and dabbing her nose with a tissue, "but some people just make good inmates."

Mr. Povenmire, the education director, didn't say anything to me, but he never had acknowledged my presence. I hoped he would drop the hard-line attitude on my last day, but I was wrong.

I walked around the inmate corridor one last time. There were as many guards here now as inmates. I waved to a case manager, a lieutenant, and the assistant warden. I even said good-bye to Mr. Flowers, who nodded and said "Good luck, Mr. White." He said my name like it tasted bad in his mouth.

I strolled past the handball courts and stopped at the breezeway where I was to meet Ella and Harry. As I waited for them to arrive, I took in the colony one more time. I breathed in the deep aroma of the banana trees. I looked hard at the sun's rays as they cut through the branches of the live oaks. I watched some inmates meander in the courtyard. Where I was headed, I didn't imagine I would see many men simply passing the time.

My mother and father waited outside. They had been divorced for almost two decades, but they came here together to greet me. I could only imagine how they felt about my prospects. A thirty-three-year-old son with massive debt, a felony conviction, no job, no home, no spouse, two children, and accumulated assets that fit into two cardboard boxes. They were worried, understandably.

And I was too. My hands were a bit shaky.

But I did feel fortunate. I had made friends with men and women I never would have known on the outside—Doc and Link, Frank Ragano and Dan Duchaine. And of course Harry and Ella. Link was right about one thing: none of us would have been friends anywhere else.

I would miss them all, but as much as anything else I would miss time. Time to daydream. Time to walk. Time to pay attention. Time to plan adventures for my new life, a new life with my children. Time to remember that *great* doesn't always mean *big*. And time, especially, with Ella.

I had no idea if I would ever see her again. I would be on federal probation for five years. I would not be allowed to leave Oxford without permission. And regardless of what Father Reynolds said, I assumed the public health authorities would not welcome ex-cons back to Carville. I didn't know if I'd ever have another conversation with Ella. I had no idea how long she would live.

I was excited, but also apprehensive. Excited about building a new home with Neil and Maggie. Overjoyed I would see them every day. Hopeful I could make up for this year apart. But I was also afraid. Afraid of going back out to a place that held so many temptations for me. Afraid I would make promises I couldn't keep. Afraid I would try to impress people with how well I would recover from failure. Afraid of returning to Oxford, where as a child I'd been scarred by a fall, and where as an adult I had acted so recklessly. I was afraid I would build new prisons for myself, the kind I had built long before I was convicted of a crime.

I heard Jimmy Harris squeeze the horn attached to his tricycle handles. He peddled toward me and stopped at the ramp.

"Good morning, young fellow," he said. I shook Jimmy's hand and I thanked him for being so generous with his stories.

"Well," he said, "I just fell in love with you as soon as we met."

I'd heard him say those very words to at least a dozen other inmates. I didn't point out that he forgot my name most of the time, but I did remind him that I was leaving.

"Good," he said, "good for you."

I wished him luck on his book.

"*King of the Microbes*," he said. "You're gonna buy one, right?"

"Absolutely."

"Good, good," he said. Jimmy squeezed his horn again and peddled off toward the patient canteen.

Ella and Harry eventually made their way down the breezeway and stopped where they always did, just inside the hallway. I stood where I always stood, just inside the inmate boundary.

"You packed?" Harry asked. I nodded. I stepped into the hallway, reached for Harry's hand, and held it in both of mine. I wanted to tell this gentle man that I was honored to be his friend, that his disfigured hand was a symbol—like a unique, broken, beautiful sculpture—that embodied something important for me that I didn't fully understand.

I wanted him to know that taking Communion with him, watching Father Reynolds place the wafer in the remnants of his palm, was a privilege and would alter, forever, how I felt about the sacrament. I wanted him to know that a tip of his hat and a smile as he rode past me in the hallways reassured me there were kind, understanding people waiting outside the gates. But I didn't say anything. As usual, Harry didn't say much either.

Then I turned to Ella. She looked so alive and vibrant. I couldn't possibly say anything adequate to this woman who had every right to be bitter and resentful, but was more content than anyone else I had ever encountered. She had come to exemplify for me what was good and pure and honest and right in all of us, an angel who'd lost her family as a girl, but made a home in a colony of outcasts. A woman whose words had directed me along a new path.

Ella always seemed to know what I needed to hear. "Any words of wisdom?" I asked.

She didn't miss a beat. "Don't forget to go to church."

I wanted them to know that they had been great examples for me. Examples of how to live a simple, fulfilling life, in spite of the facts. I wanted Ella to know that she showed me a new way to view my flaws and strengths. She even influenced how I would be a father to my children. I wanted to tell them both how much they meant to me, but I didn't.

The three of us had an imbalanced relationship. I wasn't nearly as important to Harry and Ella as they were to me. Like the doctors, nurses, and short-term patients who occupied Carville over the years, I was just another in a long line of guests. A transient passing through their secret world.

As I stepped away, down the ramp into the inmate courtyard, I didn't turn around. I walked backward, slowly. I wanted to remember everything.

I wanted to remember my good fortune. A prison sentence, any-where else, might have been lost time.

I wanted to remember the smells and the scenes. The long shadows thrown by the stately white buildings and the sweet smell of dust in the walkways that connected them. The way mist rose from the fairways of the golf course in the morning.

But most of all I wanted to remember Ella. Every detail. The way she cranked the antique handles. The way she twisted in her chair at the dance. The way she turned her disease, the most shameful known to man, into something sacred. I wanted to remember how she held her coffee mug, the way she got excited on bingo night, her smile when she said something unexpected, the joy she found in the smallest encounters, the way her skin smelled like flowers. The way she rested her hand on top of mine when I felt most alone. I wanted to remember her every word. I wanted to remember her especially whenever I was confronted with my own past, in hopes that I could face it with a fraction of her dignity.

I kept my eyes fixed on Harry and Ella. I had no idea what the future held for me. When I stepped outside the gate, I planned to stop at the edge of the levee to see which way the river flowed. After that, I didn't know.

But at some point after I settled in Oxford, I would take Ella's advice and find a church. Not just any church. A place like the church at Carville. Where the parishioners were broken and chipped and cracked. A place to go when I needed help. A place to ask forgiveness. A sacred place where people were not consumed with image or money.

I didn't know if a church like this existed, but if it did I would go. And I would pray. Not the kind of prayers I used to say for miracles or money or advancement. I would ask for something more simple. I would pray for recollection—pray that I would never forget.

I reached the bottom of the ramp and walked up the small concrete steps that led to the inmate hallway. I opened the screen door and looked at my friends again. I wanted to remember them exactly this way. Harry straddling his bicycle. Ella in her antique wheelchair.

I waved one last time, stepped inside the hallway, and turned toward freedom.

Fifteen years ago, as a prisoner, I was welcomed by the secret people. It was an honor I cherish. I left with the hope that I would never forget. But a lesson is never as clear as the moment it is learned. I have forgotten and remembered. Veered off and corrected.

But I am always drawn back. To the place where the river flows backward, where outcasts find a home, where the disfigured are beautiful. At night, I dream about the colony. Sometimes I am lost. Other times, I encounter my old friends. And sometimes, I see Ella. She glides in her chair down the empty corridors. She sways to music I cannot hear. She reminds me there is no place like home. And I know I will always be able to find her. Ella will be waiting for me. In the breeze.

September 4, 2008
Oxford, Mississippi

Ella Bounds in the breezeway.

EPILOGUE

Frank Ragano's book, *Mob Lawyer,* was published immediately after his release. He died in Tampa, Florida, in 1998. In 2007, the U.S. government released secret documents that revealed the CIA *had* paid Santo Trafficante to assassinate Fidel Castro. The report claimed all assassination attempts had failed.

After his release from prison, **Dan Duchaine** was paid an unprecedented advance for *Body Opus: Militant Weight Loss,* a book he wrote while imprisoned at Carville. He introduced DNP, the primary component of Doc's heat pill, to the bodybuilding world. Its use has been implicated in the deaths of several bodybuilders and athletes. On January 14, 2000, at age forty-eight, Dan died in his sleep of complications related to polycystic kidney disease.

In 1994, **Jefferson** moved back to New Orleans intending to open a legitimate franchise. His whereabouts are unknown.

After his release from federal custody in 1997, **Steve Read** sold yachts in Florida. He died of cancer in 1999.

I have had no contact with **Link** since he left Carville. His whereabouts are unknown.

While in prison, **Victor "Doc" Dombrowsky** was awarded three patents from the United States Patent Office. After his release in 1999, he opened a medical clinic in the Italian Alps to administer his heat pill. The procedure was internationally recognized as an effective treatment for Lyme disease and certain forms of cancer. In 2008, he was found guilty on five of eighteen federal charges related to a securities fraud scheme. He was sentenced to fourteen years in federal prison. He maintains his innocence.

Six months after I was released from federal custody, Carville residents commemorated the one hundredth anniversary of the leprosarium—and the closing of the federal prison—with a weeklong ceremony.

Jimmy Harris's self-published book, *King of the Microbes,* sold six hundred copies before he died of natural causes at age ninety-two.

Two funeral services were held for **Betty Martin,** who died on June 7, 2002. One at Carville, under the name Betty Martin, with her secret friends. The other in Baton Rouge with her prominent Louisiana family, where leprosy, Betty Martin, and her best-selling book were never mentioned.

Annie Ruth Simon remained at Carville until her passing in 2002. She is buried next to her husband in the patient cemetery.

Ella Bounds died at Carville in 1998. She is buried outside Abita Springs, Louisiana, in an unmarked grave.

In 1999, the Public Health Services closed operations at Carville. The patients were given a choice between being relocated to a hospital in Baton Rouge or living on their own and receiving a $33,000 annual stipend. **Father Reynolds** watched as the Carville residents were removed from their home, two by two, and relocated to a nondescript hospital wing in Baton Rouge. A year later, a Catholic bishop ordered Father Reynolds's transfer. He now lives in a monastery in eastern Kentucky.

Desmond "Harry" Harrington still lives at Carville. In 1999, he and thirty-six other long-term residents refused to leave their home. Today, he and fifteen others share the facility with Louisiana juvenile offenders.

My mother, **Jane Stanley,** graduated from seminary in 1998 and started a church in a poverty-stricken neighborhood in Gulfport, Mississippi, where she spends her days doing great things.

My father, **Neil White Jr.,** is a federal administrative judge in Alexandria, Louisiana.

Little Neil is a student at Princeton University.

Maggie is a student at Davidson College.

I live in Oxford, Mississippi, where I operate a small publishing company. I am married to Debbie Bell, a law professor at the University of Mississippi. We live in a log cabin in the woods. She is allergic to cologne.

In 2000, the **National Hansen's Disease Museum** was established at Carville to preserve the unique history of the residents and the leprosarium.

Although 95 percent of the population is resistant to leprosy, approximately 6,500 individuals in the United States have been diagnosed with the disease. The number of active cases requiring treatment is just over 3,000. In the United States between 150 and 200 new cases of leprosy are reported each year. Of the indigenous cases, virtually all are discovered in southern Louisiana and Texas, Gulf Coast areas where there is a high prevalence of leprosy in armadillos.

Though leprosy is now treatable, there is no vaccine; there is no test to determine who is susceptible, and the exact manner of transmission is still not known. Most leprosy patients in America never reveal the nature of their disease.

ACKNOWLEDGMENTS

I cannot begin to thank all the people who helped me with this book until I apologize to those I hurt. My deepest apology and sincere regret to Leo, George, and the fine people at Hancock Bank; to Chevis and the Peoples Bank; to my childhood friend William Weatherly; to Joe Casano, who was there to lend a hand; to my bankers, Alan and David; and to Bill Metcalf, who tried to give me a fresh start. I also apologize to a long line of investors who believed in me and lost, especially Elwood, Bill, Gerald, Margaret, William, Sherman, Ernest, Catherine, and Sarah, and to all the other lenders, vendors, advertisers, subscribers, business associates, freelance writers and photographers, and friends I hurt. And a most sincere apology to the loyal and committed employees who believed my promises. I am so sorry.

This book is not mine. It is a collective effort of hundreds of friends who took an interest in the story, including Colby Kullman, Christopher Schager, Larry Kadlec, and Cathey Riemann. Thanks for your early enthusiasm for the work. Thanks also go to the following:

To Pamela Massey, who gave me my first job out of prison. To Ken and JoAnn for giving me a place to stay until I found my footing. To Will and Patty Lewis for opening doors. To Anne Strand for listening. To Duncan for leading the Wednesday night services at the church I'd been hoping to find. To Taylor, Ollie, and St. Peter's Episcopal Church, for everything. And to the good people of Oxford for being more forgiving than I had ever imagined. There really is no place like home.

To everyone who helped with research, fact-checking, and archival images, Tanya Thomassie, Cassandra White, Kris Gilliland, Paul Johansen, Vern Evans, Ken Murphy, and Terri Fensel. To Marcia Gaudet, Tony Gould, and Zachary Gussow for writing great books about Carville and leprosy. To Dr. James Krahenbuhl, director of the National Hansen's Disease Program, who helped me better understand an extraordinarily complex disease, and Elizabeth Schexnyder, curator of the National Hansen's Disease Museum, who assisted in navigating the even more complicated culture and history of Carville. To Anwei Law and IDEA as well as José Ramirez and *The Star* for your opinions and notes on nomenclature. To all my inmate friends who were just too normal to be included in this book and, of course, to the residents of Carville for their generous storytelling.

To my great readers, Karen Bryant, Mary Ann Reed Bowen, Susannah Northart, Margaret Seicshnaydre, Linda, Julie, Tor, Priscilla, Scott, Debbie and Randy, Cheryl and Cory, Liz and Jamie, and the Oxford essayists group. To my writing teachers, Cully Randall, David Galef, Darcy Steinke, Steve Yarbrough, Romulus Linney, Marcia Norman, Tom Franklin, Dinty W. Moore, Lee Gutkind, and most especially Barry Hannah, my friend and tennis pal, who showed me the way when I finally had something to say.

To Carroll Chiles for keeping the store.

To Max and Carrie for Crow's Nest.

To my agent Jeff Kleinman of Folio Literary Management, who recognized the possibilities, waited patiently, and gently nudged me along.

To the wonderful people at William Morrow/HarperCollins for believing in this story, especially Lisa Gallagher for her enthusiasm and Laurie Chittenden for her vision, insight, and patience. The book wouldn't be the same without you. Also to Tavia Kowalchuk for your marketing brilliance, and Seale Ballenger and Ben Bruton for your invaluable publicity work. Also to Stella Connell of The Connell Agency.

To my stepmother, Jill, who was sane in the crazy years.

To Linda Peal for her forgiveness and friendship.

To my strange, wonderful, loyal mother, Jane Stanley, who is best when times are worst. To my father, Neil White Jr., a private man, for supporting me always, especially in this public confession. To Lindsay, for laughing at whatever I do and welcoming the three of us into a new family. To my wife, Debbie, a remarkable partner, editor, and friend who stepped in where Ella left off. To Little Neil and Maggie. They say parenting is a one-way street, but that's not true for us. Your innocent love while "Daddy was in camp" kept it from all falling apart.

To Judge Walter Gex for holding me accountable.

BIBLIOGRAPHY

Elwood, Julia, ed. *Carville . . . 100 Years: Carville Centennial Celebration 1894–1994*. Washington, D.C.: U.S. Department of Health and Human Services, 1994.

———. *With Love in Their Hearts: The Daughters of Charity of St. Vincent de Paul 1896–1996*. Washington, D.C.: U.S. Department of Health and Human Services, 1996.

Gaudet, Marcia. *Carville: Remembering Leprosy in America*. Jackson: University Press of Mississippi, 2004.

Gould, Tony. *A Disease Apart: Leprosy in the Modern World*. New York: St. Martin's Press, 2005.

Gussow, Zachary. *Leprosy, Racism, and Public Health: Social Policy in Chronic Disease Control*. Boulder, CO: Westview Press, 1989.

Martin, Betty. *Miracle at Carville*. New York: Doubleday, 1950.

Ramirez, José, Jr. *Squint: My Journey with Leprosy*. Jackson: University Press of Mississippi, 2009.

Secret People: The Naked Face of Leprosy in America. DVD. Directed by John Anderson and Lisa Harrison. Boston: Fanlight Productions, 1995.

Stein, Stanley, and Lawrence Blochman. *Alone No Longer: The Story of a Man Who Refused to Be One of the Living Dead*. New York: Funk & Wagnalls, 1963.

Triumph at Carville: A Tale of Leprosy in America. DVD. Directed by John Wilhelm and Sally Squires. Arlington, VA: PBS Video, 2008.

• • •

National Hansen's Disease Museum, Elizabeth Schexnyder, Curator
Physical Location: 5445 Point Clair Road, Building 12, Carville LA 70721

Mailing Address: 1770 Physicians Park Drive, Baton Rouge, LA 70816
Web: www. hrsa.gov/hansens/museum
Email: nhdpmuseum@hrsa.gov
Tel: 225-642-1950; Fax: 225-642-1949
The museum is open to the public, free of charge, Tuesdays through
Saturdays, 10:00 A.M. to 4:00 P.M. Closed federal holidays.

National Hansen's Disease Program
Mailing Address: 1770 Physicians Park Drive, Baton Rouge, LA 70816
Web: www. hrsa.gov/hansens

PHOTOGRAPH CREDITS

A portion of the author's proceeds go to IDEA, the National Hansen's Disease Museum and advocacy groups protecting the rights of persons afflicted with Hansen's disease.

IDEA
International Association for Integration Dignity and
Economic Advancement
www.idealeprosydignity.org
32 Fall St., Suite #A
P.O. Box 651
Seneca Falls, NY 13148
Tel: 315-568-5838; 888-647-4939
Email: alaw@idealeprosydignity.org

NATIONAL HANSEN'S DISEASE MUSEUM
www.hrsa.gov/hansens/museum
Physical location: 5445 Point Clair Road, Building 12
Carville, LA 70721
Mailing address: 1770 Physicians Park Drive
Baton Rouge, LA 70816
Tel: 225-642-1950
Email: nhdpmuseum@hrsa.gov